T0220494

Basic Abstract Algebra

Exercises and Solutions

Other World Scientific Titles by the Author

Introductory Topology: Exercises and Solutions
ISBN: 978-981-4583-81-7 (pbk)

Introductory Topology: Exercises and Solutions
Second Edition
ISBN: 978-981-3146-93-8
ISBN: 978-981-3148-02-4 (pbk)

An Operator Theory Problem Book
ISBN: 978-981-3236-25-7

Basic Abstract Algebra
Exercises and Solutions

Mohammed Hichem Mortad
University of Oran 1, Algeria

World Scientific

NEW JERSEY · LONDON · SINGAPORE · BEIJING · SHANGHAI · HONG KONG · TAIPEI · CHENNAI · TOKYO

Published by

World Scientific Publishing Co. Pte. Ltd.
5 Toh Tuck Link, Singapore 596224
USA office: 27 Warren Street, Suite 401-402, Hackensack, NJ 07601
UK office: 57 Shelton Street, Covent Garden, London WC2H 9HE

British Library Cataloguing-in-Publication Data
A catalogue record for this book is available from the British Library.

BASIC ABSTRACT ALGEBRA
Exercises and Solutions

Copyright © 2022 by World Scientific Publishing Co. Pte. Ltd.

All rights reserved. This book, or parts thereof, may not be reproduced in any form or by any means, electronic or mechanical, including photocopying, recording or any information storage and retrieval system now known or to be invented, without written permission from the publisher.

For photocopying of material in this volume, please pay a copying fee through the Copyright Clearance Center, Inc., 222 Rosewood Drive, Danvers, MA 01923, USA. In this case permission to photocopy is not required from the publisher.

ISBN 978-981-125-210-5 (hardcover)
ISBN 978-981-125-249-5 (paperback)
ISBN 978-981-125-211-2 (ebook for institutions)
ISBN 978-981-125-212-9 (ebook for individuals)

For any available supplementary material, please visit
https://www.worldscientific.com/worldscibooks/10.1142/12719#t=suppl

Printed in Singapore

Contents

Introduction

This manuscript aims to be an introduction to basic concepts of abstract algebra through solved exercises. Its primary audience is first (and second year in some cases) year mathematics or computer science students. Obviously, it may also be used by instructors. The book covers several topics viz. sets, logic, mappings, binary relations, groups, rings, fields, polynomials, and rational fractions.

I have been teaching this course for many years, and I have fallen back, when preparing lectures and tutorials, on many references, e.g.: [**4**], [**7**], [**11**], [**12**], [**13**], [**15**], [**18**], [**21**] and [**32**]. Some of the exercises are kindly borrowed from these references.

Abstract algebra's notions can somehow be a little hard to assimilate by students. In some sense, they are not used yet to the generalizations of certain concepts. For example, they hesitate in the first place to accept that $ab = 0$ does not necessarily imply that $a = 0$ or $b = 0$, where a and b belong to given sets. Other instances will be met below.

The way the book is structured follows to some extent other books of mine such as [**22**] and [**23**]. More precisely, each chapter in this book contains a succinct section entitled "Basics", in which readers may find the necessary definitions and results to do the exercises. In addition, this section is used for fixing notations, which might be different elsewhere. This part, however, cannot replace a detailed course on the subject. Students are also expected to have basic knowledge of notions about numbers and divisibility, among others.

Then, we have a "True or False" section. In this section, numerous questions are proposed to readers. In some cases, the questions contain traps. These questions are not meant to intimidate students, but rather to make them deepen their understanding, and to prevent them from making many silly mistakes.

The significant part of each chapter is the "Exercises with Solutions" section. This section contains a variety of exercises with very detailed solutions. Each chapter ends with the section "Supplementary Exercises".

Notice in the end that the present book is the first of a series of three books, intended for first-year students, by the same author (the other two being [**25**] and [**26**]), and to be published with "World Scientific Publishing Company". I take this opportunity to warmly thank all their staff for their patience and help (in particular, Dr. Lim Swee Cheng).

Finally, I will be happy to hear from readers about any eventual errors or suggestions at my email address: **mhmortad@gmail.com**.

Oran, October 2021
Mohammed Hichem Mortad

CHAPTER 1

Sets and Logic

1.1. Basics

1.1.1. Sets.

DEFINITION 1.1.1. A set is a well-defined collection of objects or elements. We denote sets, in general, by capital letters: A, B, C, E, F, X, Y, Z, etc. Elements or objects belonging to sets are usually designated by lowercase letters: a, b, c, x, y, z, etc.

If a is an element of A, then we write: $a \in A$. Otherwise, we write $a \notin A$.

If a set does not contain any element, it is called the empty (or void) set, and it is denoted by $\varnothing$.

DEFINITION 1.1.2. If a set has n elements, where n is a natural number, then say that the cardinal of this set is n. The cardinal of a set is denoted by "card".

A singleton is a set which has exactly one element.

DEFINITION 1.1.3. Let A and B be two sets.

(1) We say that A is a subset of B (or that B is a superset of A) if every element of A is an element of B, and we write

$$A \subset B \text{ or } B \supset A.$$

We may also say that A is contained (or included) in B (or that B contains (or includes) A). The relation "$\subset$" is called an inclusion.

(2) The two sets A and B are equal, and we write $A = B$, if

$$A \subset B \text{ and } B \subset A.$$

Otherwise, we write $A \neq B$.

We can produce new sets from old ones using known operations in mathematics.

DEFINITION 1.1.4. Let A and B be two sets.

(1) The union of A and B, denoted by $A \cup B$, is defined by

$$A \cup B = \{x : x \in A \text{ or } x \in B\}.$$

1

(2) The intersection of A and B, denoted by $A \cap B$, is defined by

$$A \cap B = \{x : \; x \in A \text{ and } x \in B\}.$$

(3) If $A \cap B = \varnothing$, then we say that the two sets are disjoint.

(4) If X is some set, and $A, B \subset X$ are such that $A \cup B = X$, and $A \cap B = \varnothing$, then we say that $\{A, B\}$ is a partition of X.

(5) The difference $A - B$ (or $A \setminus B$) is defined as

$$A - B = \{x : \; x \in A \text{ and } x \notin B\}.$$

(6) Let A be a subset of a set X. Then the set $X - A$ is called the complement of A, and it is denoted by A^c.

Remark. Students, when dealing with unions, get sometimes confused with the notion of a partition recalled above. The word "or" in the definition of the union of two sets is used in the inclusive sense. That is, any of "$x \in A$" or "$x \in B$" or "x is in both A and B" allows us to write $x \in A \cup B$. More details about the mathematical "or", in general, may be found below.

Remark. It is useful to keep in mind that:

$$x \in A^c \iff x \notin A.$$

The next easy two results are fundamental.

PROPOSITION 1.1.1. *Let A and B be two sets. Then*

$$A \subset B \iff B^c \subset A^c.$$

PROPOSITION 1.1.2. *Let A and B be two sets. Then*

$$A \cap B = \varnothing \iff A \subset B^c \iff B \subset A^c.$$

Before passing to rules of set theory, we have yet two more ways of producing new sets from old ones. Here is the first one:

DEFINITION 1.1.5. Let A and B be two sets. The cartesian product of A and B, denoted by $A \times B$, is defined by

$$A \times B = \{(a, b) : \; a \in A, \; b \in B\},$$

where (a, b) is called an ordered pair. Write $(a, b) = (c, d)$ if and only if $a = c$ and $b = d$.

If $A = B$, then we write $A \times A = A^2$.

DEFINITION 1.1.6. Let X be a set. The set of all possible subsets of X (including $\varnothing$ and X itself) is called the power set of X. We denote it by $\mathcal{P}(X)$.

1.1.2. Important sets of numbers.

- The set of positive integers or natural numbers is:

$$\mathbb{N} = \{1, 2, 3, \cdots\}.$$

The set of integers is:

$$\mathbb{Z} = \{\cdots, -2, -1, 0, 1, 2, \cdots\}.$$

The set of rational numbers (or quotients of two integers) is

$$\mathbb{Q} = \left\{\frac{m}{n} : m \in \mathbb{Z}, \ n \in \mathbb{N}\right\}.$$

A number is irrational if it is not rational. We will show, for example, that $\sqrt{2}$ is irrational. Less obviously, e and π are irrational numbers (see, e.g. [**30**] for proofs).

The set of real numbers, denoted by $\mathbb{R}$, is the union of the set of all rational and the set of all irrational numbers. So, the set of irrational numbers may be denoted by $\mathbb{R} \setminus \mathbb{Q}$. In other words, the pair of sets $\{\mathbb{Q}, \mathbb{R} \setminus \mathbb{Q}\}$ constitutes a partition of $\mathbb{R}$. Clearly

$$\mathbb{N} \subset \mathbb{Z} \subset \mathbb{Q} \subset \mathbb{R}.$$

- It is customary to write $\mathbb{Z}^*$ instead of writing $\mathbb{Z} - \{0\}$. The same remark applies to $\mathbb{Q}^*$ and $\mathbb{R}^*$. Similarly, we may define

$$\mathbb{Z}^+ = \{0, 1, 2 \cdots\}, \mathbb{Q}^+ = \left\{\frac{a}{b} : a \in \mathbb{Z}^+, b \in \mathbb{N}\right\} \text{ and } \mathbb{R}^+ = \{x \in \mathbb{R} : x \geq 0\}.$$

Also, $\mathbb{R}^*_+ = \{x \in \mathbb{R} : x > 0\}$.

- Here, we list the axioms for the real numbers:

$(\mathbb{R}, +, \cdot)$ is a commutative field (the general definition of a field will be given below) in the sense that it satisfies the following properties:

(1) If $x, y \in \mathbb{R}$, then $x + y \in \mathbb{R}$ and $x \cdot y \in \mathbb{R}$.

(2) The operations "+" and "$\cdot$" are commutative, i.e. for all $x, y \in \mathbb{R}$

$$x + y = y + x \text{ and } x \cdot y = y \cdot x.$$

(3) The operations "+" and "$\cdot$" are associative, that is

$$x + (y + z) = (x + y) + z \text{ and } x \cdot (y \cdot z) = (x \cdot y) \cdot z$$

for all $x, y, z \in \mathbb{R}$.

(4) For all $x \in \mathbb{R}$: $x + 0 = x$. We then say that 0 is the identity element with respect to "+" or merely the additive identity.

(5) Each $x \in \mathbb{R}$ possesses an additive inverse, noted $-x$, i.e.
$$x + (-x) = 0$$
for all $x \in \mathbb{R}$.

(6) For all $x \in \mathbb{R}$: $x \cdot 1 = x$. We call 1 the identity element with respect to "$\cdot$" or simply the multiplicative identity.

(7) The multiplication "$\cdot$" is distributive over the addition "$+$"
$$x \cdot (y + z) = x \cdot y + x \cdot z$$
for all $x, y, z \in \mathbb{R}$.

(8) Every element of $\mathbb{R}^* := \mathbb{R} - \{0\}$ has a multiplicative inverse, denoted by x^{-1}, i.e.
$$x \cdot x^{-1} = 1$$
for all $x \in \mathbb{R}^*$.

Remarks.

(1) All the previous axioms are satisfied by rational numbers, and so $(\mathbb{Q}, +, \cdot)$ is a commutative field.

(2) The ultimate property is not satisfied by all elements of $\mathbb{Z}$. The first seven axioms, however, are the laws of arithmetic in $\mathbb{Z}$.

(3) There is still an axiom satisfied by all elements of $\mathbb{Z}$, $\mathbb{Q}$ and $\mathbb{R}$, namely: If $ab = 0$, then $a = 0$ or $b = 0$ (zero-divisor law).

- Now, we give a basic definition of complex numbers. First, we introduce the symbol i which has the property $i^2 = -1$. A complex number is written as $x + iy$ where $x, y \in \mathbb{R}$. The set of all complex numbers is denoted by $\mathbb{C}$. Let $z = x + iy$ be a complex number. The complex conjugate of z is the complex number $x - iy$ which is denoted by $\bar{z}$. If $z, z' \in \mathbb{C}$, then
$$\overline{z + z'} = \bar{z} + \bar{z'} \text{ and } \overline{zz'} = \bar{z}\,\bar{z'}.$$

Recall in the end that every complex number z may be expressed as (the polar form of z)
$$z = re^{i\theta} = r(\cos\theta + i\sin\theta)$$
where $r = |z| = \sqrt{x^2 + y^2}$ is the modulus of z, and the angle θ is called the principal argument.

- Here, we give the definition of some class of numbers which might not be known to first-year students. Before that, recall that a polynomial is an expression of the form $a_n x^n +$

$a_{n-1}x^{n-1} + \cdots + a_1x + a_0$ where $a_0, a_1, \cdots, a_n$ are called the coefficients. The degree of a polynomial is the highest power of x associated with a non-zero coefficient.

DEFINITION 1.1.7. An algebraic number is a complex (this includes real numbers as well) root of a polynomial having rational coefficients only. The set of algebraic numbers is denoted by $\mathbb{A}$.

A non-algebraic number is called transcendental.

EXAMPLES 1.1.1. (See Exercise 1.3.37)
(1) Every rational number is algebraic.
(2) $\sqrt{2}$, $\sqrt{3}$ and $\sqrt{2} + \sqrt{3}$ are algebraic irrational numbers.
(3) $2^{p/q}$ where $p, q \in \mathbb{N}$ say, is an algebraic number.

To show a number α is algebraic, we need only find a polynomial with rational coefficients having α as a root. But, to show that β is transcendental, we must show that there is no polynomial with rational coefficients having β as its root and this is the essence of the difficulty. So, proving the transcendence of numbers can be a very difficult task. Since this is not within the scope of this book nor is it at the level of first or second-year students, we state a few results without proof, so that students will be aware of them and, besides, they will help them solve some of the exercises.

THEOREM 1.1.1. ([2]) e and π (cf. Exercise 1.3.40) are transcendental numbers.

The proof of the next result is well-documented (see, e.g. [29]).

THEOREM 1.1.2. (Lindemann-Weierstrass theorem) If $\alpha \neq 0$ is algebraic, then e^α is transcendental.

THEOREM 1.1.3. ([16]) Let α and β be algebraic numbers which differ from 0 and 1. If

$$\gamma = \frac{\ln \alpha}{\ln \beta}$$

is irrational, then it is transcendental.

Remark. There are numerous open problems in number theory. For instance, the transcendence or "algebraicity" of $e + \pi$ or $e\pi$ is still unknown.

1.1.3. Logic.

DEFINITION 1.1.8. A proposition is a sentence that is either true or false.

EXAMPLES 1.1.2.
(1) "For all real x" is not a proposition.
(2) "For all real x: $x^2 = 1$" is a proposition, and it is not true.
(3) "For all real x: $x^2 \geq 0$" is a proposition, and it is true.

DEFINITION 1.1.9. Let P and Q be two propositions.
(1) The negation of P, denoted by $\overline{P}$, is the proposition "not P". The proposition $\overline{P}$ is true when P is false.
(2) The conjunction of P and Q, noted $P \wedge Q$, is the proposition "P and Q". The proposition $P \wedge Q$ is true when P and Q are both true.
(3) The disjunction of P and Q, noted $P \vee Q$, is the proposition "P or Q". The proposition $P \vee Q$ is true when either P or Q (or both of them) are true.
(4) The implication "$P \Rightarrow Q$", read "P implies Q", is the proposition that is false if P is true and Q is false, and true in all the remaining cases.
(5) The equivalence "$P \Leftrightarrow Q$", is the double implications "$P \Rightarrow Q$" and "$Q \Rightarrow P$". It can be read "P if and only if Q". The equivalence "$P \Leftrightarrow Q$" is true in two cases only: When P and Q are both true, or when they are both false.

Remarks.

(1) The "mathematical or" may be different from everyday's "or", downright it could be the complete opposite, and this is not just in English. I have taught this course in three different languages, and some students always ask about what it means in mathematics. So, as already observed, in logic "or" means "one of them", or "both of them" (observe that we could not get rid of the "or" in the last sentence!).
(2) In logic, when one says that a "proposition is not true", it means "it is not always true". There are of course cases where a proposition is never true.
(3) An implication is a sentence like "If P, then Q".
(4) As is customary, the sentence "if and only if" may be shortened to "iff".

Now, we pass to quantifiers.

DEFINITION 1.1.10. Let $P(x)$ be some formula in $x \in A$, where A is a given set.

(1) "$\forall x \in A : P(x)$" is read "for all $x \in A$: $P(x)$". It is true if for each choice a of x, $P(a)$ is true. Otherwise, it is false, i.e. when $P(a)$ fails to be true for at least some $a \in A$. The symbol "$\forall$" is called the universal quantifier.

(2) "$\exists x \in A : P(x)$" is read "there exists at least an $x \in A$ such that $P(x)$". It is true if there is at least one $a \in A$ such that $P(a)$ is true. Otherwise, it is false, i.e. when $P(a)$ is false for all $a \in A$. The symbol "$\exists$" is called the existential quantifier.

Remarks.

(1) There are other ways of reading "$\forall x \in A$": For instance, "for every x in A", "for each x in A", etc.

(2) There are other ways of reading "$\exists x \in A$": For example, "there is an x in A", "for some x in A", etc.

How to negate quantifiers? The answer is give next:

PROPOSITION 1.1.3.

(1) *The negation of* "$\forall x \in A : P(x)$" *is* "$\exists x \in A : \overline{P(x)}$"

(2) *The negation of* "$\exists x \in A : P(x)$" *is* "$\forall x \in A : \overline{P(x)}$".

Remark. In the foregoing proposition, and when trying to show that "$\forall x \in A : P(x)$" is false, we show that "$\exists x \in A : \overline{P(x)}$" is true. If an $x \in A$ is found explicitly, then we call it a counterexample.

1.1.4. Mathematical proofs. There are several ways of proving results in mathematics, but no matter how advanced a result is, the way of proving it will use one of the following types of proofs.

- **Direct proof of** "$P \Rightarrow Q$": This the most straightforward type of proofs. How to use it? We assume P is true, then obtain via as many as needed logical deductions that the proposition Q is true.

- **Proof by contraposition of** "$P \Rightarrow Q$": Sometimes, it is not easy to see how one can use a direct proof. Since "$P \Rightarrow Q$" is always equivalent to its contrapositive "$\overline{Q} \Rightarrow \overline{P}$", to show that "$P \Rightarrow Q$" is true, we show instead that "$\overline{Q} \Rightarrow \overline{P}$" is true. How do we do that? We suppose that Q is false, and we obtain that P is false.

- **Proof by contradiction**: Given that P is true, we want to show that Q is true. How do we proceed? We assume that Q is false, then:

(1) We either obtain that P is false (in such a case, this means that the contrapositive of "$P \Rightarrow Q$" is true, and "$P \Rightarrow Q$" must be true as well);

(2) Or, we obtain contradictions with prior results.

- **Proof by cases**: This method is used when one can treat two or more cases in a way that we can exhaust all possibilities. This is better illustrated by some examples:

    (1) Suppose we want to prove a statement is true for all $n \in \mathbb{N}$. Surely, if we do that for all even numbers and for all odd numbers, then we have shown the result for all natural numbers.

    (2) If we are asked to show that a proposition is true on a set like $A = \{-1, 0, 2, 3\}$ say, then it suffices to check that the given proposition is true for -1, 0, 2 and 3.

- **Proof by induction**: This method is used when proving a proposition of the form: "for all $n \in \mathbb{N}$: $P(n)$". The method is as follows:

    (1) (Base case): We check that $P(1)$ is true (or some other integer different from 1);

    (2) (Induction step): We show the trueness of the implication $p(n) \Rightarrow p(n+1)$.

Remark. There is a second (or strong) principle of mathematical induction, and this has not been included in this manuscript. See, e.g. [**31**].

Remark. We finish this section with a welcoming word for students to the scary world of some mathematical proofs. There are the so-called constructive and non-constructive proofs, but they are not necessarily types of proofs, e.g. a non-constructive proof could be a proof by contradiction while direct proofs are, in general, constructive. However, we do not want to say a lot about it here (you better ask logicians about these concepts). Basically and naively, a constructive proof is one in which one can show the existence of some mathematical object, by providing a way of finding it. More precisely, it provides an algorithm for producing such an object. Some proofs of the Weierstrass approximation theorem (in intermediate real analysis) are constructive. A non-constructive proof only shows the existence of a mathematical object without specifying the way of finding such an example. An instance of that is, e.g. Exercise 1.3.47.

Another interesting question is: How long are mathematical proofs? The answer is surprising. Indeed, throughout history, mathematical proofs are getting bigger and bigger. These extremely long proofs may be divided into two categories.

(1) Pre-computer era: These proofs were carried out by human beings. The number of pages of these proofs can be several hundred pages, sometimes 1300 page-long or even a lot more. Reproving those results with a different method and with a much shorter proof is very much welcomed by the mathematical community.

(2) Post-computer era: Such proofs use supercalculators and if measured by the number of published journal pages, the number is just insane! They will be something like 10000 page-long proof! Apparently the longest up to 2016 is a "Two-hundred-terabyte" mathematical proof. See [**19**].

1.2. True or False

Questions. Determine, giving reasons, whether the following statements are true or false.

(1) If $C \subset A \cup B$, then either $C \subset A$ or $C \subset B$.

(2) Let E be a non-empty set and let $x \in E$. Is it correct to write $\{x\} \in E$? What about $\{x\} \subset E$?

(3) Let E be a non-empty set and let $\mathcal{P}(E)$ be its powerset. Let $x \in E$. Then it is fine to write $\{x\} \in \mathcal{P}(E)$ and $\{x\} \subset \mathcal{P}(E)$.

(4) The powerset of E, denoted by $\mathcal{P}(E)$, may well contain only one element.

(5) $\{\varnothing\}$ is the empty set.

(6) Let A and B be two non-empty sets. Then $A \times B = B \times A$.

(7) If A and B are two subsets of a given set E. If A and B have the same complement, then $A = B$.

(8) Let A and B be two sets. Then

$$A \not\subset B \Longrightarrow A \subset B^c.$$

(9) It is clear why $\mathbb{N}$ denotes the set of natural numbers. So is the case with $\mathbb{Q}$ being the set of rational numbers (the set of quotients!), and $\mathbb{R}$ with real numbers. How about $\mathbb{Z}$?

(10) The irrationality of π^2 implies that of π.

(11) Let x, y be irrational. Then xy irrational.

(12) Let x, y, z be irrational. Then xyz irrational.

(13) Let x, y be irrational. Then $x + y$ irrational.

(14) Let x, y be algebraic numbers. Then xy and $x+y$ are algebraic.

(15) Let x be a transcendental number and let y be an algebraic number. Then $x + y$ is never algebraic.

(16) At least one of $\pi + e$ and $e\pi$ is irrational.

(17) The following statement is true "$1 < 0 \Rightarrow \frac{3}{2} \in \mathbb{N}$".

(18) The negation of "$\forall x \geq 0 : 2x \geq x$" is "$\exists x < 0 : 2x < x$".

(19) Let $P(x, y)$ be some formula in x and y. Then

$$\forall x, \forall y : P(x, y) \iff \forall y, \forall x : P(x, y).$$

(20) Let $P(x, y)$ be some formula in x and y. Then

$$\exists x, \exists y : P(x, y) \iff \exists y, \exists x : P(x, y).$$

(21) Let $P(x, y)$ be some formula in x and y. Then

$$\forall x, \exists y : P(x, y) \iff \exists y, \forall x : P(x, y).$$

(22) We want to show by induction that a given statement $P(n)$ is true for all $n \in \mathbb{N}$. So it is not important to care whether $P(n_0)$ is true (for some n_0 in $\mathbb{N}$), it only suffices to show that $(P(n) \Longrightarrow P(n+1))$ is true.

Answers.

(1) False! For instance, let $A = \{0, 1\}$, $B = \{-1, 2\}$ and $C = \{1, 2\}$, then

$$C \subset A \cup B = \{-1, 0, 1, 2\},$$

but neither $C \subset A$ nor $C \subset B$!

(2) We can only write $\{x\} \subset E$. The other writing is not allowed here.

(3) Here, on the contrary, we write $\{x\} \in \mathcal{P}(E)$ since the elements of $\mathcal{P}(E)$ are sets. For the other case, we may write $\{\{x\}\} \subset \mathcal{P}(E)$ in lieu.

(4) The only case where it can contain only one element is when $E = \varnothing$, in this case we have: $\mathcal{P}(E) = \{\varnothing\}$. Otherwise, $\mathcal{P}(E)$ contains always more than one element!

(5) No! The empty set is either $\varnothing$ or $\{\}$. The set $\{\varnothing\}$ is not empty because it contains $\varnothing$ (we may write $\varnothing \in \{\varnothing\}$).

(6) No! For instance, if $A = \{0, 1\}$ and $B = \{1, 2\}$, then

$$A \times B = \{(0, 1), (0, 2), (1, 1), (1, 2)\},$$

whilst

$$B \times A = \{(1, 0), (2, 0), (1, 1), (2, 1)\} \neq A \times B.$$

(7) True! Show it!...

(8) False! For example, in $\mathbb{R}$, we have:

$$[1, 3] \not\subset [1, 2] \text{ and } [1, 3] \not\subset [1, 2]^c = (-\infty, 1) \cup (2, \infty).$$

Readers should not get confused with

$$x \in A^c \iff x \notin A.$$

It is worth recalling that:

$$A \cap B = \varnothing \implies A \subset B^c \text{ (also } B \subset A^c).$$

(9) The main reason apparently is that the German word for "numbers" is "zahlen", and this is where the notation $\mathbb{Z}$ comes from.

(10) True! Indeed, assume that π^2 is irrational. If π were rational, so would be π^2, a contradiction!

(11) False! Let $x = y = \sqrt{2}$. A proof of the irrationality of $\sqrt{2}$ will be provided in Exercise 1.3.26 below. However,

$$xy = \sqrt{2}\sqrt{2} = 2$$

is rational, i.e. it is not irrational!

(12) False! Set $x = y = z = \sqrt[3]{2}$. From Exercise 1.3.31 below, $\sqrt[3]{2}$ is irrational. However,

$$xyz = \sqrt[3]{2}\sqrt[3]{2}\sqrt[3]{2} = 2$$

is not irrational!

(13) Untrue! Take $x = \sqrt{2}$ and $y = -\sqrt{2}$. They are both irrational, and yet

$$x + y = \sqrt{2} - \sqrt{2} = 0 \in \mathbb{Q}.$$

(14) True for both statements, however, the known proofs in the literature cannot be incorporated in the present manuscript because they are a bit advanced as regards the chosen level of this book. A proof may be consulted in say [14] (Page 58).

(15) True! If $x + y$ were algebraic, it would ensue that

$$x + y - y = x$$

is algebraic, which is a contradiction! Therefore, $x + y$ is transcendant.

Remark. Readers are perhaps interested in knowing that if a and b are two transcendental numbers, then at least $a + b$ or ab is transcendental, i.e. $a+b$ and ab cannot simultaneously be algebraic. This is a well-known result which is not within the scope of the present book. A reference among many ones is [20].

(16) True. By writing

$$(x - e)(x - \pi) = x^2 - (\pi + e)x + e\pi = 0,$$

we see that if $e + \pi$ and $e\pi$ were both rational numbers, then the last equation would be of rational coefficients. This would then mean that e and π are both algebraic, and this is untrue.

(17) True! The statement is equivalent to:

$$1 \geq 0 \vee \frac{3}{2} \in \mathbb{N},$$

which is clearly true!

Remark. Logically speaking, when we start with a false statement, the whole implication is correct. For instance, $x \in \varnothing$ surely implies that I am the richest person on Earth!

(18) NO! The negation of "$\forall x \geq 0 : 2x \geq x$" IS "$\exists x \geq 0 : 2x < x$".

(19) True! The order in which the universal quantifiers occur does not matter.

(20) True! The order in which the existential quantifiers occur does not matter.

(21) False! We cannot interchange the order of two quantifiers of different kinds. See Exercise 1.3.7 for a counterexample.

(22) False! For example, let $P(n)$ be the following statement:

$$n = n + 1, \ \forall n \in \mathbb{N}.$$

It is clearly false for all $n \in \mathbb{N}$! Observe that if $P(n)$ were true, then $P(n+1)$ would be true, since $n = n + 1$ would yield $n + 1 = (n + 1) + 1 = n + 2$. Hence, it is absolutely necessary that $P(n_0)$ is true for some fixed value n_0.

1.3. Exercises with Solutions

Exercise 1.3.1. Let $A = \{\{1, 2, 3\}, \{4, 5\}, \{6, 7, 8\}\}$.

(1) What is card A?

(2) Among the following statements, indicate those which are true

$a)$ $2 \in A$; $b)$ $\{2\} \subset A$; $c)$ $\{1, 2, 3\} \in A$; $d)$ $\{4, 5\} \subset A$;

$e)$ $\{\{6, 7, 8\}\} \subset A$; $f)$ $\varnothing \in A$; $g)$ $\varnothing \subset A$.

Solution 1.3.1.

(1) A contains three elements, namely: $\{1, 2, 3\}$, $\{4, 5\}$ and $\{6, 7, 8\}$. Therefore, card $A = 3$.

(2) (a) No! since 2 is not one of the sets $\{1, 2, 3\}$, $\{4, 5\}$ and $\{6, 7, 8\}$.

(b) No, because $\{2\}$ and A are not the same type of sets.
(c) Yes!
(d) No! As mentioned in a previous question, either one can write $\{4,5\} \in A$ or $\{\{4,5\}\} \subset A$!
(e) Yes!
(f) No!
(g) Yes! The empty set is a subset of any other set!

Exercise 1.3.2. Write the powerset of E, denoted by $\mathcal{P}(E)$, in the following cases:

(1) $E = \{0\}$;
(2) $E = \{x \in \mathbb{R} : \ x^2 + 1 = 0\}$;
(3) $E = \{x \in \mathbb{R} : \ x^2 - 1 = 0\}$;
(4) $E = \{a, b, c\}$;
(5) $E = \mathbb{N}$.

Solution 1.3.2.

(1) We have:
$$\mathcal{P}(E) = \{\varnothing, \{0\}\}.$$

(2) Since the equation $x^2 + 1 = 0$ does not have any solution in $\mathbb{R}$, $E = \varnothing$ and so
$$\mathcal{P}(E) = \{\varnothing\}.$$

(3) Clearly $E = \{x \in \mathbb{R} : \ x^2 - 1 = 0\} = \{1, -1\}$, that is:
$$\mathcal{P}(E) = \{\varnothing, \{-1\}, \{1\}, \{1, -1\}\}.$$

(4) We have:
$$\mathcal{P}(E) = \{\varnothing, \{a\}, \{b\}, \{c\}, \{a, b\}, \{a, c\}, \{b, c\}, E\}.$$

(5) Practically, one cannot write all elements of $\mathcal{P}(\mathbb{N})$. We just have to remember that $\mathcal{P}(\mathbb{N})$ contains all subsets of $\mathbb{N}$!

Exercise 1.3.3. Let A, B be two sets. Show that the following statements:

(1) $A \subset B \Longleftrightarrow B^c \subset A^c$.
(2) $(A \cap B)^c = A^c \cup B^c$.
(3) $(A \cup B)^c = A^c \cap B^c$.

Solution 1.3.3. First, recall that:
$$x \in A \cup B \Longleftrightarrow x \in A \textbf{ or } x \in B,$$
$$x \in A \cap B \Longleftrightarrow x \in A \textbf{ and } x \in B.$$

So,
$$x \notin A \cup B \Longleftrightarrow x \notin A \textbf{ and } x \notin B,$$

and
$$x \notin A \cap B \iff x \notin A \text{ or } x \notin B.$$

(1) Suppose that $A \subset B$ and let us show that $B^c \subset A^c$. Let $x \in B^c$, i.e. $x \notin B$. But $A \subset B$ so $x \notin A$, i.e. $x \in A^c$.

In a similar manner, the reverse implication may be shown, and this is left to interested readers.

(2) To show the equality, we show the inclusions $(A \cap B)^c \subset A^c \cup B^c$ and $(A \cap B)^c \supset A^c \cup B^c$.

   (a) $(A \cap B)^c \subset A^c \cup B^c$: Let $x \in (A \cap B)^c$, that is $x \notin A \cap B$ so $x \notin A$ or $x \notin B$. Hence $x \in A^c$ or $x \in B^c$, and so $x \in A^c \cup B^c$.

   (b) $(A \cap B)^c \supset A^c \cup B^c$: Let $x \in A^c \cup B^c$, that is, $x \in A^c$ or $x \in B^c$. So $x \notin A$ or $x \notin B$, i.e. $x \notin A \cap B$. Thus $x \in (A \cap B)^c$.

(3) We could prove the double inclusions "$(A \cup B)^c \subset A^c \cap B^c$" and "$(A \cup B)^c \supset A^c \cap B^c$" as carried out just above. Alternatively, we show directly the equality (which could have been done in the previous question as well!). We have

$$x \in (A \cup B)^c \iff x \notin A \cup B$$
$$\iff x \notin A \text{ and } x \notin B$$
$$\iff x \in A^c \text{ and } x \in B^c$$
$$\iff x \in A^c \cap B^c.$$

Exercise 1.3.4. Let $x, y \in \mathbb{R}$, $z \in \mathbb{C}$ and let $n \in \mathbb{N}$. Put the correct sign $\Rightarrow$, $\Leftarrow$ and $\Leftrightarrow$ between the following pairs of conditions:

(1) $x^2 = 1$, $x = 1$.
(2) $n^2 = 1$, $n = 1$.
(3) $x = \pi$, $\sin x = 0$.
(4) $z = \bar{z}$, $z \in \mathbb{R}$.
(5) $x > 1$, $x \geq 2$.
(6) $n > 1$, $n \geq 2$.
(7) $x \neq y$, $x < y$.
(8) $x < y$, $y < x$.

Solution 1.3.4.

(1) $x^2 = 1 \Leftarrow x = 1$.
(2) $n^2 = 1 \Leftrightarrow n = 1$ (the implication $\Rightarrow$ holds here since n is positive).
(3) $x = \pi \Rightarrow \sin x = 0$.
(4) $z = \bar{z} \Leftrightarrow z \in \mathbb{R}$.
(5) $x > 1 \Leftarrow x \geq 2$.

(6) $n > 1 \Leftrightarrow n \geq 2$.
(7) $x \neq y \Leftarrow x < y$.

Remark. Remember that $x \neq y \Leftrightarrow$ "$x < y$ or $x > y$".

(8) There is no sign to put between the two conditions in this case!

Exercise 1.3.5.
(1) Give the negation of each of the following statements:
    (a) $\forall x \in \mathbb{R} : 3x > x$.
    (b) $\forall x \in \mathbb{R} : (x > 0) \Rightarrow (2x > x)$.
    (c) $\exists n \in \mathbb{N} \ (5n + 11 = 3n + 12)$.
    (d) $\forall \varepsilon \in \mathbb{R} : [\varepsilon > 0 \Rightarrow (\exists n \in \mathbb{N})(\frac{1}{n} < \varepsilon)]$.
(2) In the first three cases, indicate which one (s) is (are) true.

Solution 1.3.5.

(1) (a) The negation is given by

$$\exists x \in \mathbb{R} : \ 3x \leq x.$$

    (b) Recall that the negation of "$P \Rightarrow Q$", where P and Q are two propositions, is equivalent to $P \wedge \overline{Q}$ (where $\overline{Q}$ denotes the negation of Q) since "$P \Rightarrow Q$" is equivalent to the proposition $\overline{P} \vee Q$.
    Hence, the negation of the proposition is given by

$$\exists x \in \mathbb{R} : x > 0 \wedge 2x \leq x.$$

    (c) The negation is given by:

$$\forall n \in \mathbb{N} : \ 5n + 11 \neq 3n + 12$$

    (d) The negation is given by:

$$\exists \varepsilon \in \mathbb{R} : \ \varepsilon > 0 \wedge \forall n \in \mathbb{N} : \ \frac{1}{n} \geq \varepsilon.$$

(2) The first proposition is false since its negation, which is

$$\exists x = -2 : \ -6 \leq -2,$$

is patently true.

    The second proposition is true. To see why, we proceed as follows: Let $x \in \mathbb{R}$ be such that $x > 0$. Hence

$$2x = x + x > x,$$

as needed.

The third proposition is false. By solving the equation $5n + 11 = 3n + 12$, we see that $2n = 1$ or $n = 1/2$, but $1/2 \notin \mathbb{N}$. So

$$\forall n \in \mathbb{N} : \ 5n + 11 \neq 3n + 12$$

is true.

Exercise 1.3.6. Determine whether the following propositions are true or false:

(1) $\forall x \in \mathbb{R} : \ x^2 > -2$.
(2) $\exists x \in \mathbb{R} : \ x^2 > -2$.
(3) $\forall x \in \mathbb{R} : \ \sqrt{x} \geq 0$.
(4) $\exists x \in \mathbb{R} : \ \sqrt{-x^2} \leq 0$.
(5) $\forall x \in \mathbb{R}, -x^2 + x - 3 < 0$.
(6) $\exists x \in \mathbb{R} : \ x^2 - 4x + 3 = 0$.
(7) $\exists! x \in \mathbb{R} : \ x^2 - 3x + 2 = 0$.

Solution 1.3.6.

(1) True. We know that $x^2 \geq 0$ for any *real* x. In particular, $x^2 > -2$.
(2) True. Here, we are asked to give at least one real x such that $x^2 > -2$. For instance, take $x = 1$ (which is amply sufficient for the answer).
(3) False! Before comparing $\sqrt{x}$ with 0, notice that $\sqrt{x}$ is not even defined for all $x \in \mathbb{R}$. In fact, the proposition is false for all x in $(-\infty, 0)$.

   Remark. The following proposition is true:

   $$\forall x \in \mathbb{R}^+ : \sqrt{x} \geq 0.$$

(4) True. The function $x \mapsto \sqrt{-x^2}$ is defined only at $x = 0$. For this value of x, we have: $\sqrt{-0^2} = 0 \leq 0$ (observe in passing that there is not any other value of x for which the proposition is true).
(5) True. The discriminant (Δ) of $-x^2 + x - 3$ is strictly negative, so the sign of $-x^2 + x - 3$ is the same sign as that of -1, hence negative!
(6) True. The polynomial function $x^2 - 4x + 3$ admits at least one root. In fact, and is known, it has two roots which are 1 and 3.
(7) False! The question is: Does the polynomial $x^2 - 3x + 2$ have a unique root? This is obviously untrue since $x^2 - 3x + 2$ admits two roots, namely 1 and 2.

Exercise 1.3.7. Decide, giving reasons, whether the following propositions are true or false:

(1) $\forall x \in \mathbb{R}, \forall y \in \mathbb{R} : x + y = 0$.
(2) $\exists x \in \mathbb{R}, \forall y \in \mathbb{R} : x + y = 0$.
(3) $\forall x \in \mathbb{R}, \exists y \in \mathbb{R} : x + y = 0$.
(4) $\exists x \in \mathbb{R}, \exists y \in \mathbb{R} : x + y = 0$.

Solution 1.3.7.

(1) False, because
$$\exists x = 1 \in \mathbb{R}, \exists y = 0 \in \mathbb{R} : \ x + y = 1 + 0 = 1 \neq 0.$$
In other words, its negation is true.

(2) False! We cannot fix some x (*independent* of y!) for which all real y satisfy $x + y = 0$. Let us then show that the negation of the given proposition is true, i.e.
$$\forall x \in \mathbb{R}, \exists y \in \mathbb{R} : x + y \neq 0$$
is true. To see this, let x in $\mathbb{R}$. We should at least find a $y \in \mathbb{R}$ (*in this case it may depend on* x) such that $x + y \neq 0$. Just take $y = 1 - x$, then
$$(1 - x) \in \mathbb{R} \text{ and } x + y = x + 1 - x = 1 \neq 0,$$
as wished.

(3) True. Let $x \in \mathbb{R}$. Then
$$\exists y = -x \in \mathbb{R} : \ x + y = x - x = 0.$$

(4) True. It is simple in this case, we just have to explicitly give x and y such that $x + y = 0$. We take, for example, $x = -1$ and $y = 1$.

Exercise 1.3.8. Are the following propositions true

(1) $\forall x \in \mathbb{R}, \exists y \in \mathbb{R} : y = x^2$,
(2) $\forall x \in \mathbb{R}, \exists y \in \mathbb{R} : x = y^2$?

Solution 1.3.8.

(1) True. For all real x, we can always find a y in $\mathbb{R}$ such that $y = x^2$.

(2) False! Let's give a counterexample, i.e. we show that its negation is true, i.e.
$$\exists x \in \mathbb{R}, \forall y \in \mathbb{R} : x \neq y^2$$
is true. For example,
$$\exists x = -2 \in \mathbb{R}, \forall y \in \mathbb{R} : -2 = x \neq y^2.$$

Exercise 1.3.9. By explaining your answer, indicate which of following propositions is true

(1) $\exists x \in \mathbb{R}, \forall y \in \mathbb{R} : xy \geq 0$,
(2) $\forall x \in \mathbb{R}, \exists y \in \mathbb{R} : xy > 0$,
(3) $\forall x \in \mathbb{R}, \exists y \in \mathbb{R} : x + y > 0$,
(4) $\forall x \in \mathbb{R}, \exists y \in \mathbb{R} : (x + y > 0 \text{ or } x + y = 0)$,
(5) $\forall x \in \mathbb{R}, \exists y \in \mathbb{R} : (x + y > 0 \text{ and } x + y = 0)$?

Solution 1.3.9.

(1) True because

$$\exists x = 0 \in \mathbb{R}, \ \forall y \in \mathbb{R} : xy = 0 \geq 0.$$

(2) False since

$$\exists x = 0 \in \mathbb{R}, \ \forall y \in \mathbb{R} : \ xy = 0y = 0 \leq 0.$$

(3) True for

$$\forall x \in \mathbb{R}, \ \exists y = 1 - x \in \mathbb{R} : \ x + y = x + 1 - x = 1 > 0.$$

(4) True as

$$\forall x \in \mathbb{R}, \ \exists y = 1 - x \in \mathbb{R} : \ x + y = x + 1 - x = 1 > 0$$

(even if $x + y \neq 0$).
  We could also argue as follows:

$$\forall x \in \mathbb{R}, \ \exists y = -x \in \mathbb{R} : \ x + y = x - x = 0$$

(even if $x + y \not> 0$).
(5) False because we can never have $x + y > 0$ and $x + y = 0$ simultaneously.

Exercise 1.3.10. Let I be a subset of $\mathbb{R}$, and let $f : I \to \mathbb{R}$ be a function. Using quantifiers, write the following propositions symbolically, as well as their negations:

(1) f is the null function.
(2) f vanishes.
(3) f is constant.

Solution 1.3.10.

(1) f is the null function means that f is identically null, i.e. it equals zero for all values of I. In symbols,

$$\forall x \in I : f(x) = 0.$$

Its negation is therefore given by:

$$\exists x \in I : f(x) \neq 0.$$

(2) f vanishes signifies that $f(x) = 0$ for at least one point x of I. So, symbolically

$$\exists x \in I : f(x) = 0.$$

Its negation is

$$\forall x \in I : f(x) \neq 0.$$

(3) We have

$$\exists c \in \mathbb{R}, \forall x \in I : f(x) = c.$$

Its negation is:

$$\forall c \in \mathbb{R}, \exists x \in I : f(x) \neq c.$$

Exercise 1.3.11. Show that:

(1) $(\forall \varepsilon \geq 0, \; |x| \leq \varepsilon) \Rightarrow x = 0$.
(2) $(\forall \varepsilon > 0, \; |x| \leq \varepsilon) \Rightarrow x = 0$.

Solution 1.3.11.

(1) Assume that for all $\varepsilon \geq 0$: $|x| \leq \varepsilon$, and we show that $x = 0$. Since $|x| \leq \varepsilon$ holds for all $\varepsilon \geq 0$, then it particulary holds for $\varepsilon = 0$. Hence $|x| \leq 0$, i.e. $x = 0$.
(2) To show that "$(\forall \varepsilon > 0, \; |x| \leq \varepsilon) \Rightarrow x = 0$" is true, we show that its contrapositive, i.e. "$x \neq 0 \Rightarrow \exists \varepsilon > 0, \; |x| > \varepsilon$", is true. Let $x \neq 0$. There is an $\varepsilon = \frac{|x|}{2} > 0$ (since $x \neq 0$) such that $|x| > \frac{|x|}{2}$ because $x \neq 0$, and the property is shown.

Exercise 1.3.12. Explain, by considering the proposition

$$P(x) : \; x^2 \geq x, \; \forall x \geq 0,$$

why the proof by induction cannot merely be extended to the case of positive real numbers.

Solution 1.3.12. First, the proposition $P(x)$ is not true for all $x \geq 0$ since it is false for x in $(0, 1)$. Let's try to apply the reasoning by induction to this proposition anyway.

First, $P(0)$ is true because $0 \geq 0$. So, suppose that $x^2 \geq x$ and let us show that $(x+1)^2 \geq x + 1$. Since x is positive, we get

$$(x+1)^2 = x^2 + 2x + 1 \geq x + 2x + 1$$

(by the induction hypothesis). Hence

$$(x+1)^2 \geq 3x + 1 \geq x + 1.$$

So by this "pseudo-proof by induction", the proposition is true whilst we saw above that it was false! So, the verdict is: We use the proof by induction only over natural numbers.

Exercise 1.3.13. Using the proof by induction, show the following statements:

(1) $\forall n \in \mathbb{N} : 1 + 2 + \cdots + n = \frac{n(n+1)}{2}$.

(2) $\forall n \in \mathbb{N} : 1^2 + 2^2 + \cdots + n^2 = \frac{n(n+1)(2n+1)}{6}$.

(3) $\forall n \in \mathbb{N} : 1^3 + 2^3 + \cdots + n^3 = \frac{n^2(n+1)^2}{4}$.

(4) $\forall n \in \mathbb{N} : (1 + 2 + \cdots + n)^2 = 1^3 + 2^3 + \cdots + n^3$.

Solution 1.3.13.

(1) Let $P(n)$ be such a proposition. First, we check that $P(1)$ is true. Indeed, the left hand side of the equality is equal to 1 whereas the right one is $\frac{1 \times 2}{2} = 1$. Whence $P(1)$ is true.

Now, suppose that $P(n)$ is true and let us show that $P(n+1)$ is also true, i.e.

$$1 + 2 + \cdots + n + (n+1) = \frac{(n+1)(n+2)}{2}.$$

We have

$$1 + 2 + \cdots + n + (n+1) = \frac{n(n+1)}{2} + (n+1)$$
$$= (n+1)\left(\frac{n}{2} + 1\right)$$
$$= \frac{(n+1)(n+2)}{2}.$$

(2) Let $P(n)$ be this proposition. We first check that $P(1)$ is true. Indeed, the left hand side of the equality gives: $1^2 = 1$ and the right one gives $\frac{1 \times 2 \times 3}{6} = \frac{6}{6} = 1$. Hence $P(1)$ is true. Suppose that $P(n)$ is true, and let us show that $P(n+1)$ is true too, that is

$$1^2 + 2^2 + \cdots + n^2 + (n+1)^2 = \frac{(n+1)(n+2)(2n+3)}{6}.$$

We have

$$1^2 + 2^2 + \cdots + n^2 + (n+1)^2 = \frac{n(n+1)(2n+1)}{6} + (n+1)^2$$

$$= (n+1)\left(\frac{n(2n+1)}{6} + (n+1)\right)$$

$$= (n+1)\left(\frac{n(2n+1) + 6n + 6}{6}\right)$$

$$= (n+1)\left(\frac{2n^2 + 7n + 6}{6}\right)$$

$$= \frac{(n+1)(n+2)(2n+3)}{6}$$

since the two roots of $2n^2 + 7n + 6$ are -2 and $-\frac{3}{2}$.

(3) Use the same idea!

(4) We may proceed as above, or alternatively we may combine Questions 1) and 3). Indeed, let $n \in \mathbb{N}$. We then have

$$(1 + 2 + \cdots + n)^2 = \left(\frac{n(n+1)}{2}\right)^2 \quad \text{(from (1))}$$

$$= \frac{n^2(n+1)^2}{4}$$

$$= 1^3 + 2^3 + \cdots + n^3 \quad \text{(from (3))}.$$

Exercise 1.3.14. Using a proof by induction, show that the following statements hold:

(1) $4^n + 2$ is divisible by 3, for all n.

(2) For each $x > 0$ and each integer $n \geq 2$:

$$(1 + x)^n > 1 + nx.$$

Infer that $\left(1 + \frac{1}{n}\right)^n > 2$, for each $n \geq 2$.

(3) $\forall n \in \mathbb{N} : f(n) \geq n$, where f is a strictly increasing function defined from $\mathbb{N}$ into $\mathbb{N}$.

Solution 1.3.14.

(1) The statement is true for $n = 0$ because it is plain that $4^0 + 2 = 1 + 2 = 3$ is divisible by 3. Now, we suppose that $4^n + 2$ is divisible by 3, and we show that $4^{n+1} + 2$ is divisible by 3.

Since $4^n + 2$ is divisible by 3, there exists a natural integer m such that $4^n + 2 = 3m$, that is, $4^n = 3m - 2$. So

$$4^{n+1} + 2 = 4 \cdot 4^n + 2 = 4(3m - 2) + 2 = 12m - 6 = 3(4m - 2),$$

i.e. $4^{n+1} + 2$ is divisible by 3.

(2) Let $x > 0$. The statement holds for $n = 2$ because

$$(1 + x)^2 = 1 + 2x + x^2 > 1 + 2x \text{ since } x > 0.$$

Now, we suppose that $(1 + x)^n > 1 + nx$ and we show that $(1 + x)^{n+1} > 1 + (n + 1)x$. Since $1 + x > 0$, the induction assumption gives

$$(1 + x)^{n+1} = (1 + x)(1 + x)^n > (1 + x)(1 + nx).$$

But,

$$(1+x)(1+nx) = 1 + nx + x + nx^2 = 1 + (n+1)x + nx^2 > 1 + (n+1)x$$

because $nx^2 > 0$, as wished.

To answer the second claim of the question, just set $x = \frac{1}{n}$ (which is clearly strictly positive) to obtain the required inequality

$$\left(1 + \frac{1}{n}\right)^n > 1 + \frac{n}{n} = 2.$$

(3) Let $f : \mathbb{N} \to \mathbb{N}$ be a strictly increasing function. The statement is true for $n = 1$, i.e. $f(1) \geq 1$ because $f(n) \in \mathbb{N}$. Next, assume that $f(n) \geq n$ and we show that $f(n+1) \geq n+1$. By assumption $f(n) \geq n$, so

$$f(n) + 1 \geq n + 1.$$

Since f is strictly increasing, $f(n + 1) > f(n)$. Since both $f(n)$ and $f(n+1)$ are two natural integers, we get $f(n+1) \geq f(n) + 1$. Consequently $f(n + 1) \geq n + 1$, as wished.

Exercise 1.3.15. Let x be a real number such that $x \neq 1$. Show that for all $n \in \mathbb{N}$

$$1 + x + x^2 + \cdots + x^{n-1} = \frac{x^n - 1}{x - 1}.$$

Solution 1.3.15. When $n = 1$, the left hand side equals 1, and this is precisely the value of the right hand side. So, the statement is true for $n = 1$. Now, assume that

$$1 + x + x^2 + \cdots + x^{n-1} = \frac{x^n - 1}{x - 1},$$

and let us show that

$$1 + x + x^2 + \cdots + x^{n-1} + x^n = \frac{x^{n+1} - 1}{x - 1}.$$

We have

$$1 + x + x^2 + \cdots + x^{n-1} + x^n = \frac{x^n - 1}{x - 1} + x^n$$

$$= \frac{x^n - 1 + x^{n+1} - x^n}{x - 1}$$

$$= \frac{x^{n+1} - 1}{x - 1},$$

as suggested.

Exercise 1.3.16. Show that:

$$\forall n \geq 2 : \underbrace{\frac{2^{2n-1}}{n} < \binom{2n}{n} < 2^{2n-1}}_{P(n)}$$

where $\binom{p}{k} = \frac{(p)!}{k!(p-k)!}$, $p, k \in \mathbb{N}$.

Solution 1.3.16. Clearly

$$\binom{2n}{n} = \frac{(2n)!}{(2n - n)!n!} = \frac{(2n)!}{n!n!} = \frac{(2n)!}{(n!)^2}.$$

We use a proof by induction.

(1) The proposition $P(n)$ is true for $n = 2$ because

$$4 = \frac{2^{4-1}}{2} < \frac{4!}{(2!)^2} = \frac{24}{4} = 6 < 2^{4-1} = 8.$$

(2) Suppose that

$$\frac{2^{2n-1}}{n} < \frac{(2n)!}{(n!)^2} < 2^{2n-1},$$

and we show that:

$$\frac{2^{2(n+1)-1}}{n+1} < \frac{(2(n+1))!}{((n+1)!)^2} < 2^{2(n+1)-1},$$

i.e.

$$\frac{2^{2n+1}}{n+1} < \frac{(2n+2)!}{((n+1)!)^2} < 2^{2n+1}$$

(recall that

$$(2n+2)! = 1 \times 2 \times \cdots \times n \times \cdots \times 2n \times (2n+1) \times (2n+2),$$

and that $2^{2n+1} = 2^{2n} \times 2$). Denote $\frac{2^{2n-1}}{n} < \binom{2n}{n}$ by $P_1(n)$, and $\binom{2n}{n} < 2^{2n-1}$ by $P_2(n)$.

Let's start by proving that $P_1(n)$ is true. By hypothesis, we have

$$\frac{2^{2n-1}}{n} < \frac{(2n)!}{(n!)^2},$$

then

$$\frac{(2n+1)(2n+2)}{(n+1)(n+1)} \times \frac{2^{2n-1}}{n} < \frac{(2n)!}{(n!)^2} \times \frac{(2n+1)(2n+2)}{(n+1)(n+1)},$$

(because $\frac{(2n+1)(2n+2)}{(n+1)(n+1)} \geq 0$). After simplification we get

$$\frac{(2n+1)2^{2n}}{n(n+1)} < \frac{(2n+2))!}{((n+1)!)^2}.$$

So, to show that $P_1(n+1)$ is true, it suffices to prove that

$$\frac{2^{2n+1}}{n+1} < \frac{(2n+1)2^{2n}}{n(n+1)}.$$

But

$$\frac{2^{2n+1}}{n+1} < \frac{(2n+1)2^{2n}}{n(n+1)}$$

$$\Longleftrightarrow 2^{2n+1} < \frac{(2n+1)2^{2n}}{n}$$

$$\Longleftrightarrow 2 < \frac{2n+1}{n}$$

$$\Longleftrightarrow 2n < 2n+1$$

$$\Longleftrightarrow 0 < 1.$$

So, we have shown that $P_1(n)$ is true. To prove that $P_2(n)$ is true, we adopt a similar method. We get

$$\frac{(2n)!}{(n!)^2} < 2^{2n-1} \implies \frac{(2n+1)(2n+2)}{(n+1)(n+1)} \times \frac{(2n)!}{(n!)^2} < 2^{2n-1} \times \frac{(2n+1)(2n+2)}{(n+1)(n+1)}.$$

Hence

$$\frac{(2n+2)!}{((n+1)!)^2} < 2^{2n} \times \frac{2n+1}{n+1}.$$

We will be done as soon as we check that $2^{2n} \times \frac{2n+1}{n+1} < 2^{2n+1}$.
But

$$2^{2n} \times \frac{2n+1}{n+1} < 2^{2n+1} \iff \frac{2n+1}{n+1} < 2 \iff 2n+1 < 2n+2 \iff 1 < 2.$$

This marks the end of the proof.

Exercise 1.3.17. Let E be a set such $\operatorname{card} E = n$, where $n \in \mathbb{N}$. Let $\mathcal{P}(E)$ be the powerset of E. Show that $\operatorname{card} \mathcal{P}(E) = 2^n$.

Solution 1.3.17. We provide two methods.

(1) We use a proof by induction. The proposition is true for $n = 1$. Indeed, if E is a singleton, for example, $E = \{x\}$, then obviously

$$\mathcal{P}(E) = \{\varnothing, E\} \text{ so } \operatorname{card} \mathcal{P}(E) = 2 = 2^1.$$

Now suppose that if the set E of n elements, then its powerset has 2^n elements. Then we show that if a set has $n+1$ elements, then its power set has 2^{n+1} elements.

Set $F = \{x_1, x_2, \cdots, x_n, x_{n+1}\}$. The powerset in this case consists of the powerset of $\{x_1, x_2, \cdots, x_n\}$, as well as sets obtained by taking the union of each element of the power set and the element x_{n+1} in each union. By hypothesis, the number of subsets of E is 2^n. From what has just been explained, the cardinal of $\{x_1, x_2, \cdots, x_n, x_{n+1}\}$ will be exactly twice the cardinal of $\{x_1, x_2, \cdots, x_n\}$. In other words,

$$\operatorname{card} \mathcal{P}(F) = 2 \times 2^n = 2^{n+1},$$

completing the proof.

(2) First, remember that a combination of p elements of a set E is a subset of E which has p elements. Now, one of the fundamental properties is that in a set of n elements, the number of combinations of p elements is given by:

$$\binom{n}{p} = \frac{n!}{p!(n-p)!}.$$

On the other hand, we know that $\mathcal{P}(E)$ contains the parts having: 0 elements (we only have one here that is the empty set), 1 element, 2 elements, $\cdots$, n elements. So we get:

$$\operatorname{card} \mathcal{P}(E) = \binom{n}{0} + \binom{n}{1} + \cdots + \binom{n}{n}.$$

A moment's thought allows us to notice that the previous expression is noting but the binomial formula of $(1+1)^n = 2^n$. Thus

$$\operatorname{card} \mathcal{P}(E) = (1+1)^n = 2^n,$$

as suggested.

Exercise 1.3.18. Is the statement: "For all $n \in \mathbb{N}$, $n^2 + n + 41$ is a prime number" true?

Solution 1.3.18. One could be tempted to use a proof by induction. The statement is true for $n = 1$, as in this case $n^2 + n + 41 = 43$ is a prime number. In the meantime, the second step of induction seems not to lead anywhere! If we carry on with the checking, we see that the statement is still true for $n = 2, 3, 4, 5, 6, 7, 8, 9, 10, 11, 12, 13$. Shall we dare and write "etc."? Beginners could say "and so on", as they are convinced that the statement is surely true, they just don't know yet how to prove it! If they check more cases, they see that the statement is still true for all $n = 14, 15, 16, \cdots, 39$. Now, they say: No way! This has got to be true! The mathematical rigor is now obliged to intervene. The statement is wrong for $n = 40$ for

$$40^2 + 40 + 41 = 1681 = 41 \times 41$$

is obviously not prime! The main lesson students should learn from this exercise is: *A statement is true only when a correct proof is provided.*

Exercise 1.3.19. Show that the difference of two even numbers remains even.

Solution 1.3.19. If $n, m \in \mathbb{Z}$ are even, then $a = 2k$ and $b = 2l$ for some $k, l \in \mathbb{Z}$. Thus

$$n - m = 2k - 2l = 2(k - l)$$

is an even integer for $k - l \in \mathbb{Z}$.

When $n, m \in \mathbb{N}$ are even, the same proof remains valid and $n - m$ is even if one further assumes that $n \geq m$.

Exercise 1.3.20. Show that among any arbitrary choice of three positive integers, we can always pick two whose sum is even.

Solution 1.3.20. We use a proof by cases. If we are given three natural numbers, then the only possible cases are (we may even reduce these four cases to two only):

(1) All three integers are even.
(2) Two of them are even, and the remaining number is odd.
(3) One of them is even, and the other two are odd.
(4) All three integers are odd.

It becomes therefore clear that in each of the preceding cases, we may always find two integers having an even sum, and this completes the proof.

Exercise 1.3.21. Show that the product of two consecutive integers is always even.

Solution 1.3.21. Let m be a natural number. We aim to show that $m(m + 1)$ is even. We use a proof by cases.

(1) If m is even, then $m(m+1)$ is obviously even.
(2) If m is odd, then $m+1$ is even, as is $m(m+1)$.

Exercise 1.3.22. Let $n \in \mathbb{N}$. Show that if $(n^2 - 1)$ is not divisible by 8, then n is even.

Solution 1.3.22. Let's show that the contrapositive of the implication is true. Assume that n is not even, i.e. n is odd, and so there is a $k \in \mathbb{N}$ such that $n = 2k + 1$. Hence

$$n^2 - 1 = (2k+1)^2 - 1 = 4k^2 + 4k + 1 - 1 = 4k^2 + 4k = 4k(k+1).$$

According to Exercise 1.3.21, $k(k+1)$ is even, which means that we can write it as $2p$, with $p \in \mathbb{N}$. Whence

$$n^2 - 1 = 4k(k+1) = 8p,$$

showing that $n^2 - 1$ is divisible by 8.

Exercise 1.3.23. Let $n \in \mathbb{N}$. Prove that if n^2 is even, then n is even too.

Solution 1.3.23. There are different ways to answer this question. The commonly known proof is the one in which we show that the contrapositive is true. So let's show that if n is not even, i.e. if it is odd, then n^2 is not even, that is, it is odd. If n is odd, then it may be written as $2k + 1$ for some k in $\mathbb{N}$, by the definition of an odd integer. So

$$n^2 = (2k+1)^2 = 4k^2 + 4k + 1 = 2\underbrace{(2k^2 + 2k)}_{\in \mathbb{N}} + 1,$$

hence n^2 is odd, as expected.

Can we prove this statement using a direct proof? The answer is positive. To this end, assume that n^2 is even. Since $n^2 + n = n(n+1)$, $n^2 + n$ is even by Exercise 1.3.21. Since $n^2 + n \in \mathbb{N}$ and $n^2 + n \geq n^2$, it follows that $n^2 + n - n^2 = n$ is even by Exercise 1.3.19.

Remark. If n is even, i.e. $n = 2k$ for some $k \in \mathbb{N}$, then

$$n^2 = (2k)^2 = 4k^2 = 2(\underbrace{2k^2}_{k \in \mathbb{N}})$$

is even too. Therefore, if $n \in \mathbb{N}$ (or even in $\mathbb{Z}$), then

$$n \text{ is even} \iff n^2 \text{ is even}.$$

Equivalently,

$$n \text{ is odd} \iff n^2 \text{ is odd}.$$

Exercise 1.3.24. Show that if n^2 is a multiple of 3, then so is n, where $n \in \mathbb{N}$.

Solution 1.3.24. We are showing that the contrapositive of the given implication is true. So, assume that n is not a multiple of 3, whereby it is necessarily written as either $3k + 1$ or $3k + 2$ with $k \in \mathbb{N}$. We will then have

$$n^2 = (3k + 1)^2 = 9k^2 + 6k + 1 = 3 \underbrace{(3k^2 + 2k)}_{\text{natural number}} + 1$$

while in the second case, we get

$$n^2 = (3k + 2)^2 = 9k^2 + 12k + 4 = 3 \underbrace{(3k^2 + 4k + 1)}_{\text{natural number}} + 1.$$

So, in both cases, n^2 is not a multiple of 3. Accordingly, we have established that if n^2 is a multiple of 3, then n too is a multiple of 3.

Exercise 1.3.25. Let $n \in \mathbb{N}$. Prove that if n^3 is even, then n is even too.

Solution 1.3.25. The proof is fairly similar to that of Exercise 1.3.23. Assume that n is odd, i.e. $n = 2k + 1$ for a certain $k \in \mathbb{N}$. Then

$$n^3 = (2k + 1)^3 = 8k^3 + 12k^2 + 6k + 1 = 2(4k^3 + 6k^2 + 3k) + 1 = 2l + 1$$

is odd since $l = 4k^3 + 6k^2 + 3k \in \mathbb{N}$.

Exercise 1.3.26. (See Exercise 1.3.46 for another proof) Show that $\sqrt{2}$ is not rational.

Solution 1.3.26. We will use a proof by contradiction. Suppose, on the contrary, that $\sqrt{2}$ is rational, i.e. there are two natural integers n and m such that $\sqrt{2} = \frac{m}{n}$. Assume that n and m are co-prime.

We have $m^2 = 2n^2$, i.e. m^2 is even, as is m according to Exercise 1.3.23. Therefore there is a natural number k such that $m = 2k$. The equation $m^2 = 2n^2$ then becomes $n^2 = 2k^2$. Similarly, we may show that n is necessarily even. Thus $\gcd(m, n) \geq 2$, contradicting the hypothesis that m and n are relatively prime.

Exercise 1.3.27. (See Exercise 1.3.46 for another proof) Prove that $\sqrt{3}$ is not rational.

Solution 1.3.27. To show that $\sqrt{3}$ is irrational and as in the previous solution, assume there are two coprime natural integers n and m obeying $\sqrt{3} = \frac{m}{n}$. Then $m^2 = 3n^2$, so that m^2 is multiple of 3. But, according to Exercise 1.3.24, m is a multiple of 3, i.e. $m = 3k$ with $k \in \mathbb{N}$. Consequently, the equation $m^2 = 3n^2$ becomes $3k^2 = n^2$. The same idea applied to n implies that it is a multiple of 3 too. Then

$\gcd(m,n) \geq 3$ and so $\gcd(m,n) \neq 1$, which is is a contradiction for n and m are taken to be relatively prime. Accordingly, $\sqrt{3}$ is irrational, as wished.

Exercise 1.3.28. Show that

(1) $\sqrt{2} + \sqrt{3} \notin \mathbb{Q}$;

(2) $\frac{\ln 2}{\ln 3} \notin \mathbb{Q}$;

(3) $0.336433643364 \cdots \in \mathbb{Q}$.

Solution 1.3.28.

(1) We give two methods.

- Suppose that $\sqrt{3} + \sqrt{2}$ is rational. Since

$$(\sqrt{3} + \sqrt{2})(\sqrt{3} - \sqrt{2}) = 1,$$

$\sqrt{3} - \sqrt{2}$ would be rational, and

$$\sqrt{3} + \sqrt{2} + \sqrt{3} - \sqrt{2} = 2\sqrt{3}$$

would be rational as well. This is, however, impossible according to Exercise 1.3.27.

- Now, we give the second method. Assume that $\sqrt{3} + \sqrt{2}$ is rational, that is $\sqrt{3} + \sqrt{2} = r$ where $r \in \mathbb{Q}$. Then

$$\sqrt{3} = r - \sqrt{2} \implies 3 = r^2 - 2r\sqrt{2} + 2 \implies \sqrt{2} = \frac{r^2 - 1}{2r} \in \mathbb{Q},$$

which is untrue because $\sqrt{2}$ is rational.

(2) Suppose that $\frac{\ln 2}{\ln 3}$ is rational, that is $\frac{\ln 2}{\ln 3} = \frac{p}{q}$ where p and q are two natural coprime integers. Then

$$\frac{\ln 2}{\ln 3} = \frac{p}{q} \implies q \ln 2 = p \ln 3 \implies \ln 2^q = \ln 3^p \implies 2^q = 3^p$$

and so we ended up with a contradiction because 2^q is even while 3^p is odd.

(3) Set $m = 0.336433643364 \cdots$. Then

$$10^4 m = 3364.33643364 \cdots \text{ so } 10^4 m - m = 9999m = 3364.$$

Therefore

$$m = 0.336433643364 \cdots = \frac{3364}{9999}, \text{ i.e. } m \text{ is rational.}$$

Exercise 1.3.29. (See Exercise 1.3.46 for another proof) Let p be a prime number. Prove that $\sqrt{p}$ is not rational.

Hint: Use the following standard result: If a, b are two non integers and if p is a prime number dividing ab, then p divides a or p divides b.

Solution 1.3.29. Suppose that

$$\sqrt{p} = \frac{a}{b}, \ a \in \mathbb{N}, b \in \mathbb{N} \text{ with } \gcd(a, b) = 1.$$

So

$$pb^2 = a^2.$$

Then p divides a^2. Using the hint, we know that p divides a as p is a prime number. We can write $a = pk$ with k a natural integer. Replacing a by pk in the equation above, we get $b^2 = pk^2$. The same argument gives that p divides b. So $\gcd(a, b)$ is not equal to 1, which is absurd! Thus $\sqrt{p}$ is irrational.

Exercise 1.3.30. (See Exercise 1.3.46 for another proof) Show that $\sqrt{6}$ is irrational.

Solution 1.3.30. Since 6 is not a prime number, we cannot, alas, use Exercise 1.3.29. So, we give the following proof by contradiction combined with a proof by cases. Assume that $\sqrt{6}$ is rational, then

$$\sqrt{6} = \frac{a}{b}, \ a \in \mathbb{N}, b \in \mathbb{N} \text{ with } \gcd(a, b) = 1.$$

Hence

$$6b^2 = a^2.$$

Four possibilities are investigated.

(1) If a and b are both even, then $\gcd(a, b) \geq 2$, whereby $\gcd(a, b) \neq 1$, which is absurd.
(2) If a is odd and b is even, then a^2 is odd and $6b^2$ is even! That is $6b^2 \neq a^2$, leading to a contradiction.
(3) If a and b are both odd, then a^2 is odd and $6b^2$ becomes even! In other words $6b^2 \neq a^2$, which is absurd.
(4) If a is even and b is odd, then we can write

$$a = 2p \text{ and } b = 2q + 1 \text{ with } p, q \in \mathbb{N}.$$

Hence

$$24q^2 + 24q + 6 = 4p^2 \text{ or } 12q^2 + 12q + 3 = 2p^2.$$

But $12q^2 + 12q + 3$ is odd because it can be written as $2(6q^2 + 6q + 1) + 1$ where $6q^2 + 6q + 1 \in \mathbb{N}$, while $2p^2$ is obviously even. This is again a contradiction.

Consequently, $\sqrt{6}$ is not rational.

Exercise 1.3.31. (See Exercise 1.4.13 for another proof) Show that $\sqrt[3]{2}$ is not rational. What about $\sqrt[4]{2}$?

Solution 1.3.31. We may proceed as in the case of square roots. Assume $\sqrt[3]{2}$ is rational in lowest terms, i.e. write $\sqrt[3]{2} = p/q$, where $p, q \in \mathbb{N}$ and $\gcd(p, q) = 1$. Now,

$$\sqrt[3]{2} = \frac{p}{q} \implies 2 = \frac{p^3}{q^3} \implies p^3 = 2q^3.$$

So, p^3 would be even, as would p be by Exercise 1.3.24, i.e. $p = 2k$ for some $k \in \mathbb{N}$. The above equation would then become $8k^3 = 2q^3$ or merely $q^3 = 4k^3$. In particular, q^3 would be even leading to the "evenness" of q. This is the sought contradiction. Thus, $\sqrt[3]{2}$ is irrational.

In the end, $\sqrt[4]{2}$ is not rational either. Observe that $\sqrt[4]{2} = \sqrt{\sqrt{2}}$. If we write $\sqrt[4]{2} = p/q$, where p and q are two natural numbers, then $\sqrt{2} = p^2/q^2$ which is impossible as we already know that $\sqrt{2}$ is irrational.

Exercise 1.3.32. Let $x \in \mathbb{Q}^*$. Prove that if $x^2 \in \mathbb{N}$, then $x \in \mathbb{Z}^*$.

Solution 1.3.32. Since x is rational, we have

$$x = \frac{a}{b}, \ a \in \mathbb{N}, b \in \mathbb{Z}^* \text{ and } \gcd(a, b) = 1.$$

Hence $x^2 = a^2/b^2$. Since a and b are coprime, so are a^2 and b^2. Observe that b^2 divides a^2 as $b^2 x^2 = a^2$. Since $x^2 \in \mathbb{N}$, $b^2 = 1$, and hence $b = \pm 1$ so $x = \pm a$, i.e. x belongs to $\mathbb{Z}^*$.

Exercise 1.3.33. Let x be an irrational number. Let n, m be two integers. Show that $\frac{nx}{m}$ is irrational too. Is $\sqrt{8}$ a rational number?

Solution 1.3.33. Assume that $\frac{nx}{m}$ is rational, i.e.

$$\frac{nx}{m} = \frac{p}{q},$$

where p and q are relatively prime integers. So

$$x = \frac{mp}{nq},$$

i.e. x would be rational, contradicting the irrationality of x. Hence $\frac{nx}{m}$ must be rational.

As for the second question, write $\sqrt{8} = 2\sqrt{2}$, then from Exercise 1.3.26, we already know that $\sqrt{2}$ is irrational. Since 2 is rational, the first part of the solution then allows us to infer that $2\sqrt{2}$ or $\sqrt{8}$ is irrational.

Exercise 1.3.34. Let $x \in \mathbb{R}^+$. Do we have

$$\sqrt{x} \text{ irrational} \implies \sqrt[4]{x} \text{ irrational?}$$

What about

$$\sqrt[4]{x} \text{ irrational} \implies \sqrt[3]{x} \text{ irrational?}$$

Solution 1.3.34. The first implication is true. Let us show that its contrapositive is true in lieu. If $\sqrt[4]{x}$ is rational, then $\sqrt[4]{x} = \frac{a}{b}$ where $a, b \in \mathbb{N}$. So $\sqrt{x}$ is also rational because

$$\sqrt{x} = \frac{a^2}{b^2}$$

with $a^2, b^2 \in \mathbb{N}$.

The second implication is wrong. For a counterexample, $\sqrt[4]{8}$ is irrational (readers are asked to show that), whilst $\sqrt[3]{8} = 2$ is rational.

Exercise 1.3.35. Let r be a rational number and let x be an irrational number. Show that $r + x$ is irrational.

Solution 1.3.35. Write $r = p/q$, where $p \in \mathbb{Z}$ and $q \in \mathbb{N}$. We use a proof by contradiction, and so let us assume, on the contrary, that $r + x$ is rational. So, for some $n \in \mathbb{Z}$ and $m \in \mathbb{N}$, we may write

$$r + x = \frac{n}{m} \implies x = \frac{n}{m} - r = \frac{n}{m} - \frac{p}{q} = \frac{qn - pm}{qm},$$

i.e. x would be rational, contradicting the assumption. Thus, $r + x$ is irrational.

Exercise 1.3.36.

(1) Is the sum of two irrationals always irrational?
(2) Show that if x and y are irrationals, then $x + y$ and $x - y$ cannot simultaneously be rational.

Solution 1.3.36.

(1) The answer is negative. For example, $\sqrt{2}$ is irrational, as is $1 - \sqrt{2}$. However,

$$\sqrt{2} + 1 - \sqrt{2} = 1 \notin \mathbb{R} \setminus \mathbb{Q}.$$

(2) Let x, y be both irrational numbers. Assume that

$$x + y = \frac{p}{q} \text{ and } x - y = \frac{r}{s}$$

where p, q, r, s are integers with $q \neq 0$ and $s \neq 0$. Hence

$$2x = \frac{p}{q} + \frac{r}{s} \text{ and } 2y = \frac{p}{q} - \frac{r}{s}$$

so that

$$x = \frac{ps + qr}{2qs} \text{ and } y = \frac{ps - qr}{2qs}.$$

Thus, x and y would both be rational numbers, and this is a contradiction. Thus, either $x + y$ or $x - y$ is irrational.

Exercise 1.3.37. Show the following statements:
(1) Every rational number is algebraic.
(2) $\sqrt{2}$, $\sqrt{3}$ and $\sqrt{2} + \sqrt{3}$ are algebraic irrational numbers.
(3) $\sqrt{2} + \sqrt[3]{5}$ is an algebraic number.
(4) $2^{q/p}$ where $p, q \in \mathbb{N}$ say, is an algebraic number.

Solution 1.3.37.
(1) Let p/q be a rational number. Then, $qx - p = 0$ is an equation with rational numbers, and whose solution is p/q.
(2) The irrationality of $\sqrt{2}$, $\sqrt{3}$ and $\sqrt{2} + \sqrt{3}$ has already been established.
  (a) $\sqrt{2}$ is an algebraic number for it is a solution of $x^2 - 2 = 0$ (which has rational coefficients).
  (b) $\sqrt{3}$ is an algebraic number as it is a solution of $x^2 - 3 = 0$.
  (c) To find a polynomial with rational coefficients having $\sqrt{2}+\sqrt{3}$ as a root, set $x = \sqrt{2}+\sqrt{3}$. Then $x^2 = 2+3+2\sqrt{6}$ and so $x^2 - 5 = 2\sqrt{6}$ (which still a non-algebraic equation). Therefore,

$$(x^2 - 5)^2 = 24 \text{ or } x^4 - 10x^2 + 1 = 0.$$

  Since the last equation is satisfied by $\sqrt{2} + \sqrt{3}$, we now know that $\sqrt{2} + \sqrt{3}$ is an algebraic number.
(3) The same idea applies here with slightly longer calculations. Put $x = \sqrt{2} + \sqrt[3]{5}$. Then $x - \sqrt{2} = \sqrt[3]{5}$ and so

$$5 = (x - \sqrt{2})^3 = x^3 - 3x^2\sqrt{2} + 6x - 2\sqrt{2} = x^3 + 6x - (3x^2 + 2)\sqrt{2},$$

  i.e.

$$x^3 + 6x - 5 = (3x^2 + 2)\sqrt{2} \text{ or } (x^3 + 6x - 5)^2 = 2(3x^2 + 2)^2.$$

  Whence

$$x^6 - 6x^4 - 10x^3 + 12x^2 - 60x + 17 = 0$$

  is an algebraic equation having $x = \sqrt{2} + \sqrt[3]{5}$ as one of its solutions.
(4) $2^{q/p}$ is easily seen to be a solution of the equation $x^p - 2^q = 0$ whose coefficients are rational numbers, thereby $2^{q/p}$ is algebraic, as needed.

Exercise 1.3.38. Show that $\ln 2$ is irrational.

Solution 1.3.38. For the sake of contradiction, assume that $\ln 2$ is rational, i.e. $\ln 2 = p/q$, where $p, q \in \mathbb{N}$. Then

$$\frac{q}{p} \ln 2 = 1 \Longrightarrow \ln 2^{q/p} = 1 \Longrightarrow 2^{q/p} = e^{\ln 2^{q/p}} = e.$$

This is manifestly a contradiction for e is transcendental (by Theorem 1.1.1) while $2^{q/p}$ is algebraic (Exercise 1.3.37). Consequently, $\ln 2$ is not rational.

Remark. It would interesting to see a transcendence-free proof of the irrationality of $\ln 2$.

Exercise 1.3.39. Let n be in $\mathbb{N}$, $n \geq 2$. Show that e^n is not algebraic. In particular, e^n is irrational.

Solution 1.3.39. Assume that e^n is algebraic. That is, e^n is a root of some polynomial $p(x)$ of degree m say. Hence $p(e^n) = 0$. By setting $q(x) = p(x^n)$, we see that q becomes a polynomial of degree nm. Obviously,

$$p(e^n) = q(e) = 0.$$

So, e would be algebraic, and this contradicts Theorem 1.1.1.

Exercise 1.3.40. Using Theorem 1.1.2, show that π is transcendental.

Solution 1.3.40. If π were algebraic, so would be $2i\pi$. Hence $e^{2i\pi}$ would be transcendental by Theorem 1.1.2. However, $e^{2i\pi} = 1$ is an integer, and so there is no way it can be transcendental! Therefore, π is transcendental, as wished.

Exercise 1.3.41. Show that $\ln \alpha$ is transcendental, where α is algebraic, $\alpha > 0$ and $\alpha \neq 1$. What about $(\ln \alpha)^n$, where $n \in \mathbb{N}$?

Solution 1.3.41. Aiming for a contradiction, suppose $\ln \alpha$ is algebraic, i.e. $\ln \alpha = \beta$, where β is algebraic. Hence $\alpha = e^\beta$ would be transcendental by Theorem 1.1.2. This is a contradiction since α is algebraic. Thus, $\ln \alpha$ is transcendental.

Regarding the second question, we argue as in Exercise 1.3.39. If $(\ln \alpha)^n$ were algebraic, it would be a root of some polynomial $p(x)$ of degree m say. So $p[(\ln \alpha)^n] = 0$. By setting $q(x) = p(x^n)$, we see that q is a polynomial of degree nm. Obviously,

$$p[(\ln \alpha)^n] = q(\ln \alpha) = 0.$$

So, $\ln \alpha$ would be algebraic, and this contradicts the first part of the solution.

Exercise 1.3.42. Show that $\ln 2 / \ln 3$ and $2^{\sqrt{2}}$ are transcendental.

Solution 1.3.42. By Exercise 1.3.28, $\ln 2 / \ln 3$ is irrational, and so Theorem 1.1.3 gives the transcendence of $\ln 2 / \ln 3$.

To show that $2^{\sqrt{2}}$ is transcendental, and for the sake of contradiction, assume that $2^{\sqrt{2}}$ is algebraic. It is seen that

$$\frac{\ln 2^{\sqrt{2}}}{\ln 2} = \frac{\sqrt{2} \ln 2}{\ln 2} = \sqrt{2}$$

which is not rational. So, Theorem 1.1.3 would imply that $\sqrt{2}$ is transcendental, and this is absurd. Thus, $2^{\sqrt{2}}$ is not algebraic, as needed.

Exercise 1.3.43. Let a, b and c be three integers such that

$$a^2 + b^2 = c^2.$$

Show that either a or b is even.

Remark. This result says that if a, b, c represent the lengths of the sides of a right triangle, where c is the length of the side opposite the right angle, and all lengths are in $\mathbb{N}$, the either a or b is even. This is a reformulation of the exercise using the Pythagorean theorem.

Solution 1.3.43. We use a proof by contradiction. Suppose that none of a and b is even, i.e. assume that they are both odd. Hence a^2 and b^2 are odd. That is

$$\exists n, m \in \mathbb{Z} : a = 2n + 1 \text{ and } b = 2m + 1,$$

so

$$a^2 + b^2 = 4n^2 + 4n + 1 + 4m^2 + 4m + 1 = 4 \underbrace{\left(n^2 + n + m^2 + m\right)}_{= p \in \mathbb{N}} + 2.$$

Hence c^2 is even, and by Exercise 1.3.23, c is also even, i.e. $c = 2q$ with q an integer. Finally, we get

$$c^2 = 4q^2 = 4p + 2,$$

which leads to

$$q^2 - p = \frac{1}{2},$$

and that is absurd since $q^2 - p$ is an integer while $\frac{1}{2}$ is not.

Exercise 1.3.44. Show that the sum of two odd perfect squares is never a perfect square.

Solution 1.3.44. Let a and b two odd perfect squares, i.e. $a = n^2$ and $b = m^2$, and they are odd numbers. We need to show that $a + b$ is not a perfect square. So, assume $a+b$ is a perfect square, i.e. $a+b = p^2$ ($p \in \mathbb{N}$), then we try to get a contradiction.

First, observe that $a + b$, being the sum of two odd numbers, is necessarily even, and so p^2 is even. Since n^2 and m^2 are odd, so are n and m. Also, as p^2 is even, so is p. Put differently, $n = 2k + 1$, $m = 2k' + 1$ and $p = 2l$, where $k, k', l \in \mathbb{N}$.

On expansion,

$$a + b = p^2 \iff (2k + 1)^2 + (2k' + 1)^2 = (2l)^2$$
$$\iff 4k^2 + 4k + 1 + 4k'^2 + 4k' + 1 = 4l^2$$

and so $4(k^2 + k + k'^2 + k') + 2 = 4l^2$. Hence

$$2\underbrace{(k^2 + k + k'^2 + k')}_{\in \mathbb{N}} + 1 = 2l^2.$$

The right side of the previous equation is even whereas the left side is odd, and this is impossible. Accordingly, $a + b$ is not a perfect square.

Exercise 1.3.45. (Cf. [9]) Let a, b, c be odd integers. Show that the equation $ax^2 + bx + c = 0$ has no integer solution.

Solution 1.3.45. Let a, b, c be odd integers, and assume that the equation $ax^2 + bx + c = 0$ does have an integer solution, noted n. So, n is either even or odd. Since a, b and c are odd, $a = 2k + 1$, $b = 2l + 1$ and $c = 2m + 1$ for certain $k, l, m \in \mathbb{Z}$.

If n is even, i.e. $n = 2p$ where $p \in \mathbb{Z}$, the equation $ax^2 + bx + c = 0$ then becomes

$$(2k + 1)(2p)^2 + (2l + 1)2p + 2m + 1 = 0$$

or

$$2[(2k + 1)2p^2 + (2l + 1)p + m] + 1 = 0,$$

which is a flagrant absurdity as the left side is an odd number whilst the right side is an even number.

When n is odd, a similar argument may be applied. Alternatively, there is a simpler way (a similar idea could have been used in the first step of the proof anyway). It reads: If n satisfies the equation $ax^2 + bx + c = 0$, then $an^2 + bn + c = 0$. Since n is odd, so is n^2. Since the product of two odd numbers is odd, an^2 and bn are both odd. Their sum $an^2 + bn$, however, is even. Therefore, $an^2 + bn + c$ is an odd number, and so it cannot be equal to any even number.

To conclude, the equation $ax^2 + bx + c = 0$ cannot have an integer solution, whenever the coefficients a, b and c are odd integers.

Exercise 1.3.46. (Rational root test, cf. [9], cf. Exercise 1.4.13) Let a, b, c be integers such that $a, c \neq 0$, and let $f(x) = ax^2 + bx + c$, $x \in \mathbb{R}$.

(1) Show that if the equation $f(x) = 0$ admits a rational solution, denoted by $r = p/q$ (with $\gcd(p, q) = 1$), then p must divide c and q must divide a.

(2) Infer the irrationality of say: $\sqrt{2}$, $\sqrt{3}$, $\sqrt{s}$ (where s is a prime number) and $\sqrt{3/2}$.

Solution 1.3.46.

(1) ([9]) Since $r = p/q$ is a solution of $f(x) = 0$, $f(r) = 0$. So, $ap^2/q^2 + bp/q + c = 0$. Multiplying the previous equation by q^2 gives

$$cq^2 + bpq + ap^2 = 0 \text{ or } ap^2 = -cq^2 - bpq.$$

Since q divides the right side, it divides the left side, i.e. it divides ap^2. Since $\gcd(p, q) = 1$, q must divide a.

Now, let's go back the last displayed equation, and write it as

$$cq^2 = -bpq - ap^2.$$

Since p divides the right side, it divides the left side, i.e. it divides cq^2. As $\gcd(p, q) = 1$, it is seen that p must divide c, and this completes the proof.

(2) Let us apply the previous result to establish the irrationality of the given numbers.

(a) For the irrationality of $\sqrt{2}$, consider $f(x) = x^2 - 2$. Now, assume that this polynomial has a rational root denoted by r, and written as p/q in lowest terms. By the above result, p must divide -2, and q must divide 1. The options are $p = \pm 1, \pm 2$, and $q = \pm 1$. So, $r = \pm 1, \pm 2$. However, clearly none of $1, -1, 2, -2$ satisfies $x^2 - 2 = 0$. Thus, $x^2 - 2 = 0$ does not possess any rational root, and so all other roots, if there are any, are irrational. Since $\sqrt{2}$ is a solution of $x^2 - 2 = 0$, it has to be be irrational, and this is enough for us.

(b) For the irrationality of $\sqrt{3}$, consider $f(x) = x^2 - 3$. As before, if the preceding polynomial had rational roots, they would be $3, -3, 1, -1$. Since they do not satisfy the given equation, the solutions of $x^2 - 3 = 0$ are irrational. In particular, $\sqrt{3}$ is not rational.

(c) Let s be a prime number, then consider $x^2 - s = 0$. The only possible rational solutions would be $s, -s, 1, -1$, and none of the latter satisfies $x^2 - s = 0$, whereby $\sqrt{s}$ is necessarily irrational.

(d) Consider in the end the equation $2x^2 - 3 = 0$. If it had rational solutions $r = p/q$, then p would divide 3 and q would divide 2. So, $p = \pm 3, \pm 1$ and $q = \pm 2, \pm 1$. So, $r = \pm 3/2, \pm 1/2, \pm 3, \pm 1$. Again, none of these numbers is a solution of $2x^2 - 3 = 0$. So, if there is any other solution, it must be irrational. Since $\sqrt{3/2}$ is clearly a solution of $2x^2 - 3 = 0$, it must therefore be irrational, as needed.

Exercise 1.3.47. Show the following result, specifying the type of proof used: "There are two irrational numbers a and b such that a^b is rational."

Hint: Try with $\sqrt{2}^{\sqrt{2}}$.

Solution 1.3.47. We use a (non-constructive) proof by cases.

(1) First case: If $\sqrt{2}^{\sqrt{2}}$ is rational, then we take $a = b = \sqrt{2}$ (which are both irrational!).

(2) Second case: If $\sqrt{2}^{\sqrt{2}}$ is irrational, then we take $a = \sqrt{2}^{\sqrt{2}}$ and $b = \sqrt{2}$, and they are both irrational (a being irrational by hypothesis). So we get

$$a^b = \left(\sqrt{2}^{\sqrt{2}}\right)^{\sqrt{2}} = \sqrt{2}^{(\sqrt{2})^2} = \sqrt{2}^2 = 2,$$

which is clearly rational, and that marks the end of the proof.

Remark. The way we have proceeded with the solution was only to illustrate the use of proofs by cases. A priori, it does not say which is true! However, readers are asked in Exercise 1.4.12 to show that $\sqrt{2}^{\sqrt{2}}$ is in fact irrational. Nonetheless, students at this level are not expected to come up with a proof of the irrationality of $\sqrt{2}^{\sqrt{2}}$. So, they should not be too disappointed should they fail to find a proof on their own.

Remark. There is a simpler way of answering this question: We already know that e and $\ln 2$ are both irrational numbers, and

$$e^{\ln 2} = 2$$

is a natural number!

1.4. Supplementary Exercises

Exercise 1.4.1. Let A and B be two sets. Show that the following propositions are equivalent:

(1) $A \subset B$;
(2) $A \cup B = B$;
(3) $A \cap B = A$.

Exercise 1.4.2. Let A and B be two sets. Prove that:
$$A \cup B = A \cap B \iff A = B.$$

Exercise 1.4.3. Let X and Y be two sets. Show that
$$\mathcal{P}(X) \cap \mathcal{P}(X) = \mathcal{P}(X \cap Y)$$
and
$$\mathcal{P}(X) \cup \mathcal{P}(X) \subset \mathcal{P}(X \cup Y),$$
where $\mathcal{P}$ denotes the powerset.

Exercise 1.4.4. Determine whether the following propositions are true:

(1) $\exists x \in \mathbb{R}, \ \forall y \in \mathbb{R} : xy > 0$.
(2) $\forall x \in \mathbb{R}, \ \exists y \in \mathbb{R} : \ x - y > 0$.
(3)

$[(\forall x \in \mathbb{R}, \ \exists y \in \mathbb{R} : \ x + y > 0) \text{ and } (\forall x \in \mathbb{R}, \ \exists y \in \mathbb{R} : \ x + y = 0)$.

Exercise 1.4.5. Is the following proposition true
$$[(1 > 2 \implies 3 = 4) \implies (0 \neq 1)]?$$

Exercise 1.4.6. Let $f : \mathbb{R} \to \mathbb{R}$ be a function. Express symbolically (using quantifiers) the following statements:

(1) f is bounded from above.
(2) f never takes the same value twice.
(3) f is increasing.

Exercise 1.4.7. Prove that:
$$\forall n \in \mathbb{N} : \ -1 + 2 - 3 + \cdots + (-1)^n n = \frac{(-1)^n (2n + 1) - 1}{4}.$$

Exercise 1.4.8. Let $k \in \mathbb{N}$ be given. Show that if $n \in \mathbb{N}$ and n^k is even, then n is also even.

Exercise 1.4.9. Let $k \in \mathbb{N}$ be given. Is $\sqrt[k]{2}$ rational?

Exercise 1.4.10. Let $a \geq 2$. Prove that $\sqrt{a}$ is irrational if a is not a perfect square (i.e. $a \neq n^2$, $n \in \mathbb{N} - \{1\}$).

Exercise 1.4.11. Let $p \in \mathbb{N}$. Prove that $\sqrt{2} + \sqrt{p}$ is irrational.

Exercise 1.4.12. (The proof is quite advanced) (Try to) Show that $\sqrt{2}^{\sqrt{2}}$ is irrational.

Exercise 1.4.13. (Rational root test, see, e.g. [**9**]) Let $a_0, a_1, \cdots, a_n$ be integers, where $a_0, a_n \neq 0$, and let

$$f(x) = a_n x^n + \cdots + a_1 x + a_0 = 0, \ x \in \mathbb{R}$$

(1) Show that if the equation $f(x) = 0$ admits a rational solution, denoted by $r = p/q$ (with $\gcd(p, q) = 1$), then p must divide a_0 and q must divide a_n.
(2) By choosing appropriate polynomials, deduce that, e.g. $\sqrt[3]{2}$, $\sqrt[4]{3}$ and $\sqrt[5]{6}$ are irrational numbers.

Exercise 1.4.14. Show that $\sqrt{2} + \sqrt[3]{2}$ is irrational.

CHAPTER 2

Mappings

2.1. Basics

We give the most basic and naive definition of a function.

DEFINITION 2.1.1. Let X and Y be two sets. A mapping f (or function or map) from X to Y is a rule which assigns to each element $x \in X$ a unique element $y \in Y$. We generally express this by writing $y = f(x)$.

The set X is called the domain of f and Y is called the codomain of f.

DEFINITION 2.1.2. Two functions $f : X \to Y$ and $g : X \to Z$ are said to be equal if:

$$f(x) = g(x), \ \forall x \in X.$$

DEFINITION 2.1.3. The identity function on a set X is the function $f : X \to X$ defined by $f(x) = x$. It is usually denoted by id_X.

DEFINITION 2.1.4. Let $f : X \to Y$ be a function and let $A \subset X$. The function $g : A \to Y$ is called a restriction of f to A if

$$f(x) = g(x), \ \forall x \in A.$$

We may denote g by f_A and we may say: f restricted to A.

We now give two important subsets of the domain and the codomain of a given function.

DEFINITION 2.1.5. Let $f : X \to Y$ be a function. Let $A \subset X$ and $B \subset Y$. Then:

(1) The image (or range) of A, denoted by $f(A)$, is defined to be

$$f(A) = \{f(x) : \ x \in A\}.$$

(2) The inverse image (or the preimage) of B, denoted by $f^{-1}(B)$, is defined by

$$f^{-1}(B) = \{x \in X : \ f(x) \in B\}.$$

Remarks.

(1) Readers should always remember that the image $f(A)$ is a subset of Y whereas $f^{-1}(B)$ is a subset of X.

(2) Also

$$x \in f^{-1}(B) \Longleftrightarrow f(x) \in B$$

and

$$y \in f(A) \Longleftrightarrow \exists x \in A : \ f(x) = y.$$

(3) Readers should bear in mind that, in general, $f(X) \subsetneq Y$.

(4) The inverse image is defined for any function f, i.e. f is not assumed to be invertible (in the sense of Definition 2.1.7 below), even though in the case of invertibility we still denote the inverse by f^{-1} (in such case the two notations coincide).

DEFINITION 2.1.6. Let X, Y and Z be three non-empty sets. Let $f : X \to Y$ and $g : Y \to Z$ be two functions. The composition (or composite) of g with f is the function denoted by $g \circ f$, and defined from X into Z by

$$(g \circ f)(x) = g(f(x)).$$

Remark. It is worth noticing that "$\circ$" is not commutative but it is associative. It is also easy to see that for all f

$$f \circ \mathrm{id} = \mathrm{id} \circ f = f$$

where id is the identity function on X. The next natural question is: When is a function invertible? We state this definition separately.

DEFINITION 2.1.7. A function $f : X \to Y$ is invertible, if there is a function $g : Y \to X$ such that:

$$f \circ g = \mathrm{id}_Y \text{ and } g \circ f = \mathrm{id}_X,$$

where id_X and id_Y are the identity functions on X and Y respectively. In other words, f is invertible if

$$(f \circ g)(y) = y \text{ and } (f \circ g)(x) = x$$

for all $x \in X$ and all $y \in Y$. The function g is called the inverse function of f, and it is denoted by f^{-1}.

Next, we introduce concepts which help for the checking of the invertibility of functions.

DEFINITION 2.1.8. Let $f : X \to Y$ be a function. We say that f is injective (or one-to-one) if:

$$\forall x, x' \in X : \ (x \neq x' \Longrightarrow f(x) \neq f(x'))$$

or equivalently:

$$\forall x, x' \in X : \ (f(x) = f(x') \Longrightarrow x = x').$$

We have just noticed that $f(X)$ does not always equal Y. But when it does, then this has a particular appellation.

DEFINITION 2.1.9. Let $f : X \to Y$ be a function. We say that f is surjective (or onto) if:

$$\forall y \in Y, \ \exists x \in X : \ f(x) = y,$$

or equivalently if $f(X) = Y$.

Remark. The word "onto" may be a little bit misleading for some readers who have not taken their first course on functions in English. They might think that a function from X onto Y is just like a function from X into Y! This is of course not true as we have just said that: onto $=$ surjective!

It can happen that a function is injective and surjective at the same time (as in the last example). This leads to the following fundamental notion.

DEFINITION 2.1.10. Let $f : X \to Y$ be a function. We say that f is bijective (or one-to-one correspondence) if f is injective and surjective simultaneously, that is, iff

$$\forall y \in Y, \ \exists! x \in X : \ f(x) = y$$

(recall that $\exists!$ means "there is a unique").

One primary use of bijective functions is invertibility.

THEOREM 2.1.1. *A function is invertible if and only if it is bijective.*

2.2. True or False

Questions. Determine, giving reasons, whether the following statements are true or false.

(1) Let $f : \mathbb{R} \to \mathbb{R}$ be a function defined by $f(x) = 2x + 1$. Which is true
   (a) $f^{-1}(1) = 0$
   (b) or $f^{-1}(\{1\}) = 0$?
(2) Let E and F be two sets. Let $f : E \to F$ be a mapping. If $A \subset E$, then $f(A)$ is never empty unless $A = \varnothing$.
(3) Let E and F be two sets. Let $f : E \to F$ be a mapping. Then it may happen that $f^{-1}(B) = \varnothing$ even if $B \neq \varnothing$ (where $B \subset F$).
(4) Let E and F be two sets. Let $f : E \to F$ be a mapping. Then $f(E) = F$.

(5) Let E and F be two sets. Let $f : E \to F$ be a mapping. Then $f^{-1}(F) = E$.

(6) Write the definition of injectivity and surjectivity of a map defined from $\mathbb{R}^3$ onto $\mathbb{R}^2$.

(7) The function $x \mapsto x^2$ is injective but not surjective!

(8) Let E be a finite set, and let $f : E \to E$ be a certain function. Then f is either bijective, **or** it is neither injective nor surjective.

(9) Let $f : E \to F$ be a map. Then

$$\forall x, x' \in E : (x = x' \Rightarrow f(x) = f(x')) \Longrightarrow f \text{ is injective.}$$

(10) Describe geometrically the injectivity of a function defined from $\mathbb{R}$ into $\mathbb{R}$.

(11) Describe geometrically the surjectivity of a function defined from $\mathbb{R}$ into $\mathbb{R}$.

(12) Describe geometrically the bijectivity of a function defined from $\mathbb{R}$ into $\mathbb{R}$.

Answers.

(1) We write $f^{-1}(1) = 0$. For the other case, we may write $f^{-1}(\{1\}) = \{0\}$.

(2) True. This follows easily from the definition.

(3) False! We will see a counterexample in say Exercise 2.3.5. Readers might be interested in knowing that if $f : E \to F$ is a function and $B \subset f(E)$, then

$$f^{-1}(B) \neq \varnothing \Longleftrightarrow B \neq \varnothing.$$

(4) False! This only occurs when f is surjective. For a counterexample, any non-surjective function will do.

(5) True. Indeed,

$$f^{-1}(F) = \{x \in E : f(x) \in F\} = E.$$

(6) Let $f : \mathbb{R}^3 \to \mathbb{R}^2$ be a function. Then f is injective if

$$\forall (x, y, z), (x', y', z') \in \mathbb{R}^3 :$$

$$[f(x, y, z) = f(x', y', z') \Longrightarrow (x, y, z) = (x', y', z')],$$

and f is surjective if:

$$\forall (u, v) \in \mathbb{R}^2 : \exists (x, y, z) \in \mathbb{R}^3 : f(x, y, z) = (u, v).$$

(7) It is primordial to know the domain ("starting") and the image ("arrival") set! An expression as $x \mapsto x^2$ may not even define a function. If the function is well-defined, Exercise 2.3.7 gives

us four examples of "this expression" which yield in each case a different answer.

(8) Amazingly, that is true! This is just a reformulation of a quite interesting result, namely: If $f : E \to E$ is some function where E is a finite set, then

$$f \text{ injective } \Longleftrightarrow f \text{ surjective } \Longleftrightarrow f \text{ bijective.}$$

A proof may be consulted in Exercise 2.3.24. Notice that the assumption of the finiteness of E is primordial. Indeed, in Exercise 2.3.7 we have a surjective function that is not injective, and another example of an injective function that is not surjective.

(9) No! This property holds for *all* functions!!

(10) Let Γ be the graph of f. Then f is injective if and only if each line parallel to the real axis intersects Γ at most once.

(11) Let Γ be the graph of f. Then f is surjective if and only if each line parallel to the real axis intersects Γ at least once.

(12) Let Γ be the graph of f. Then f is bijective if and only if each line parallel to the real axis intersects Γ exactly once.

2.3. Exercises with Solutions

Exercise 2.3.1. Explain why the following expressions are not well-defined functions:

(1) $f : \mathbb{R} \to \mathbb{R}$, $x \mapsto f(x) = \begin{cases} 1 - x, & x \leq 1, \\ 1 + x, & x \geq 1; \end{cases}$

(2) $f : \mathbb{R}^2 \to \mathbb{R}$, $(x, y) \mapsto f(x, y) = x^y$;

(3) $f : \mathbb{Q} \to \mathbb{Z}$, $x = \frac{m}{n} \mapsto f(x) = n$ where n and m are two integers;

(4) $f : \mathbb{R} \setminus \mathbb{Q} \times \mathbb{R} \setminus \mathbb{Q} \to \mathbb{R} \setminus \mathbb{Q}$, $(x, y) \mapsto f(x, y) = x + y$.

Solution 2.3.1.

(1) f is not well-defined because $f(1)$ admits two *different* values (that are 0 and 2)!

(2) f is not well-defined because, and as is known, the function $t \mapsto a^t$ is defined for $a > 0$. For example, what is the value of $(-1)^{1/2}$?!

(3) f is not well-defined because $\frac{1}{2} = \frac{2}{4}$ but we cannot decide whether

$$f\left(\frac{1}{2}\right) = 2 \text{ or } f\left(\frac{1}{2}\right) = f\left(\frac{2}{4}\right) = 4!$$

(4) For example, let $x = \sqrt{3} \in \mathbb{R} \setminus \mathbb{Q}$ and $y = -\sqrt{3} \in \mathbb{R} \setminus \mathbb{Q}$. Then $f(x, y) = \sqrt{3} - \sqrt{3} = 0 \notin \mathbb{R} \setminus \mathbb{Q}$.

Exercise 2.3.2. Let E and F be two sets, and let $f : E \to F$ be a function.

(1) Prove that

$$A \subset B \Longrightarrow f(A) \subset f(B),$$

where $A, B \subset E$.

(2) Show that

$$A \subset B \Longrightarrow f^{-1}(A) \subset f^{-1}(B),$$

where $A, B \subset F$.

Solution 2.3.2.

(1) Assume that $A \subset B$ and let us show that $f(A) \subset f(B)$. Let $y \in f(A)$. By the definition of the image of a function, there is at least an x in A such that $y = f(x)$. Since $A \subset B$, we have $y = f(x) \in f(B)$, as desired.

(2) Suppose that $A \subset B$ and let us show that $f^{-1}(A) \subset f^{-1}(B)$. Let $x \in f^{-1}(A)$. By definition of the inverse range, $f(x) \in A$, and so $f(x) \in B$ as $A \subset B$. Hence $x \in f^{-1}(B)$.

Exercise 2.3.3. Let E and F be two sets and let $f : E \to F$ be a map. Let $A_1, A_2 \subset E$ and $B, B_1, B_2 \subset F$.

(1) Show that $f(A_1 \cup A_2) = f(A_1) \cup f(A_2)$.

(2) (a) Show that $f(A_1 \cap A_2) \subset f(A_1) \cap f(A_2)$.

    (b) Give an example showing that the equality does not always hold.

    (c) Show that

    f is injective if and only if $f(A_1 \cap A_2) = f(A_1) \cap f(A_2)$,

        for all A_1 and A_2.

(3) Show that $f^{-1}(B_1 \cup B_2) = f^{-1}(B_1) \cup f^{-1}(B_2)$.

(4) Show that $f^{-1}(B_1 \cap B_2) = f(B_1) \cap f(B_2)$.

(5) Show that $f^{-1}(B^c) = (f^{-1}(B))^c$ where "c" denotes the complement.

Solution 2.3.3.

(1) We can show the equality directly, but we choose to show the two inclusions.

(a) We have

$$y \in f(A_1 \cup A_2) \implies y = f(x) \text{ for some } x \in A_1 \cup A_2$$
$$\implies y = f(x) \text{ for some } x \in A_1 \text{ or } y = f(x) \text{ for some } x \in A_2$$
$$\implies y \in f(A_1) \cup f(A_2).$$

(b) We have

$$y \in f(A_1) \cup f(A_2) \implies y \in f(A_1) \text{ or } y \in f(A_2)$$
$$\implies y = f(x), \ x \in A_1 \text{ or } y = f(x), \ x \in A_2$$
$$\implies y = f(x), \ x \in A_1 \cup A_2$$
$$\implies y \in f(A_1 \cup A_2).$$

(2) (a) Clearly

$$A_1 \cap A_2 \subset A_1 \text{ and } A_1 \cap A_2 \subset A_2.$$

Hence

$$f(A_1 \cap A_2) \subset f(A_1) \text{ and } f(A_1 \cap A_2) \subset f(A_2),$$

that is,

$$f(A_1 \cap A_2) \subset f(A_1) \cap f(A_2).$$

(b) For the counterexample, take the function f defined from $\mathbb{R}$ into $\mathbb{R}$ by $f(x) = x^2$, then set $A = (-\infty, 0]$ and $B = [0, \infty)$. Hence

$$f(A \cap B) = f(\{0\}) = \{0\} \neq [0, \infty) = f(A) \cap f(B).$$

(c) Assume that f is injective. As seen above, we always have $f(A_1 \cap A_2) \subset f(A_1) \cap f(A_2)$ (which is true for *any* function f).

Now, we show the other inclusion. Let $y \in f(A_1) \cap f(A_2)$ and so $y \in f(A_1)$ *and* $y \in f(A_2)$. Hence $f(x_1) = y$ and $f(x_2) = y$ for some $x_1 \in A_1$ and $x_2 \in A_2$. Therefore, $f(x_1) = f(x_2)$ $(= y)$. Since f is one-to-one, $x_1 = x_2$. Thus, $x_1 = x_2 \in A_1 \cap A_2$, thereby $y \in f(A_1 \cap A_2)$, as wished.

To show the other implication, assume that $f(A_1 \cap A_2) = f(A_1) \cap f(A_2)$ for all sets A_1 and A_2, and let us show that f is injective.

Let x_1, x_2 be such that $f(x_1) = f(x_2)$, and assume that $x_1 \neq x_2$. So $\{x_1\} \cap \{x_2\} = \emptyset$. By hypothesis, we would get $f(\{x_1\} \cap \{x_2\}) = f(\{x_1\}) \cap f(\{x_2\})$, which is not true for

$$f(\{x_1\} \cap \{x_2\}) = f(\varnothing) = \varnothing \neq f(\{x_1\}) \cap f(\{x_2\}) = f(\{x_1\}) \ (= f(\{x_2\})).$$

Thus, necessarily $x_1 = x_2$, i.e. we have shown the injectivity of f, as needed.

(3) We have:

$$x \in f^{-1}(B_1 \cup B_2) \iff f(x) \in B_1 \cup B_2$$
$$\iff f(x) \in B_1 \text{ or } f(x) \in B_2$$
$$\iff x \in f^{-1}(B_1) \text{ or } x \in f^{-1}(B_2)$$
$$\iff x \in f^{-1}(B_1) \cup f^{-1}(B_2)$$

(4) Similarly, we have:

$$x \in f^{-1}(B_1 \cap B_2) \iff f(x) \in B_1 \cap B_2$$
$$\iff f(x) \in B_1 \text{ and } f(x) \in B_2$$
$$\iff x \in f^{-1}(B_1) \text{ and } x \in f^{-1}(B_2)$$
$$\iff x \in f^{-1}(B_1) \cap f^{-1}(B_2)$$

(5) Clearly

$$x \in f^{-1}(B^c) \iff f(x) \in B^c$$
$$\iff f(x) \notin B$$
$$\iff x \notin f^{-1}(B)$$
$$\iff x \in (f^{-1}(B))^c.$$

Exercise 2.3.4. Let $f : E \to F$ and $g : F \to G$ be two functions. Let $A \subset F$, $B \subset E$ and $C \subset G$.

(1) Show that
$$(g \circ f)^{-1}(C) = f^{-1}(g^{-1}(C)).$$

(2) Show that $f(f^{-1}(A)) \subset A$. Does equality always hold?

(3) Show that $B \subset f^{-1}(f(B))$. Is the equality always true?

(4) Show that f is surjective iff $f(f^{-1}(A)) = A$ for all A.

(5) Show that f is injective iff $B = f^{-1}(f(B))$ for all B.

Solution 2.3.4.

(1) We have

$$x \in (g \circ f)^{-1}(C) \Longleftrightarrow (g \circ f)(x) \in C$$
$$\Longleftrightarrow g(f(x)) \in C$$
$$\Longleftrightarrow f(x) \in g^{-1}(C)$$
$$\Longleftrightarrow x \in f^{-1}(g^{-1}(C)).$$

(2) We have

$$y \in f(f^{-1}(A)) \Longrightarrow \exists x \in f^{-1}(A) : \ y = f(x) \Longrightarrow y = f(x) \in A.$$

For the counterexample, let $f : \{0,1\} \to \{0,1\}$ be defined by $f(0) = 0$ and $f(1) = 0$. If $A = \{0,1\}$, we have:

$$f^{-1}(A) = \{x \in \{0,1\} : \ f(x) = 0 \in \{0,1\}\} = A,$$

but

$$f(f^{-1}(A)) = \{0\} \neq A.$$

(3) The same method as that of the previous answer can be applied.

For a counterexample, let $f : \{0,1\} \to \{0,1\}$ be defined by $f(0) = 0$ and $f(1) = 0$. If $B = \{0\}$, we have $f(B) = \{0\}$, and so

$$f^{-1}(f(B)) = f^{-1}(\{0\}) = \{0,1\} \neq B,$$

as needed.

(4) Suppose that f is surjective. We always have $f(f^{-1}(A)) \subset A$ (for *any* f). Let's show the other inclusion. Let $y \in A$. Since f is surjective, there is an $x \in X$ such that $f(x) = y$ or $x \in f^{-1}(A)$. So $y = f(x) \in f(f^{-1}(A))$ and

$$A \subset f(f^{-1}(A)) \subset A \Longrightarrow A = f(f^{-1}(A)).$$

Conversely, let's show that f is surjective. Let $y \in F$. Since $\{y\} \subset F$, we get $f(f^{-1}(\{y\})) = \{y\}$. Thus $f^{-1}(\{y\})$ is never empty, that is, it contains at least one element, i.e. for some $x \in E : x \in f^{-1}(\{y\})$ or $f(x) = y$. Hence f is surjective.

(5) Assume that f is injective, and we show that $f^{-1}(f(B)) \subset B$. Let $x \in f^{-1}(f(B))$ and so $f(x) \in f(B)$. This means that there is a certain $x' \in B$ such that $f(x) = f(x')$. However, f is injective, and so $x = x' \in B$.

Let's now show that f is injective, and suppose that $B = f^{-1}(f(B))$ for any set B. Let $x, x' \in B$ be such that $f(x) = f(x')$. If $B = \{x\}$, then $f(x') \in f(B)$ so $x' \in f^{-1}(f(B)) = B = \{x\}$. Hence $x = x'$, i.e. f is injective.

Exercise 2.3.5. Let $f : \mathbb{R} \to \mathbb{R}$ be the function defined by
$$f(x) = x^2, \ \forall x \in \mathbb{R}.$$
Find:
$$f([-1,1]), \ f([-1,2]), \ f^{-1}(\{1\}), \ f^{-1}(\{-1\}), \ f^{-1}([0,4]),$$
$$f^{-1}([-2,4]) \text{ and } f^{-1}([-2,1]).$$

Solution 2.3.5.

(1) We have:
$$f([-1,1]) = \{f(x) : x \in [-1,1]\} = \{x^2 : x \in [-1,1]\}.$$
So, we need to find all the numbers of the form x^2 where $-1 \le x \le 1$. It is easy to see that if $-1 \le x \le 1$, then $0 \le x^2 \le 1$. Hence
$$f([-1,1]) = [0,1].$$

(2) Notice that $[-1,2] = [-1,1] \cup [1,2]$, so that by Exercise 2.3.3, we obtain
$$f([-1,2]) = f([-1,1] \cup [1,2]) = f([-1,1]) \cup f([1,2]).$$
So, it only remains to find $f([1,2])$. We have:
$$f([1,2]) = \{x^2 : x \in [1,2]\} = [1,4].$$
Hence
$$f([-1,2]) = f([-1,1]) \cup f([1,2]) = [0,1] \cup [1,4] = [0,4].$$

(3) We have:
$$f^{-1}(\{1\}) = \{x \in \mathbb{R} : x^2 = 1\}.$$
The real solutions of $x^2 = 1$ are $x = \pm 1$, and so
$$f^{-1}(\{1\}) = \{1, -1\}.$$

(4) It is clear that
$$f^{-1}(\{-1\}) = \{x \in \mathbb{R} : x^2 = -1\} = \varnothing$$
since $x^2 = -1$ does not possess any real solution.

(5) By definition $f^{-1}([0,4]) = \{x \in \mathbb{R} : x^2 \in [0,4]\}$. In this case, we must look for all x such that $0 \le x^2 \le 4$. The safest method here perhaps is to take the positive square root in the previous inequalities (since all terms are positive!). We get $0 \le \sqrt{x^2} = |x| \le 2$, that is $-2 \le x \le 2$. Therefore,
$$f^{-1}([0,4]) = [-2,2].$$

(6) Write

$$f^{-1}([-2,4]) = f^{-1}([-2,0) \cup [0,4]) = f^{-1}([-2,0)) \cup f^{-1}([0,4]).$$

But $f^{-1}([-2,0)) = \varnothing$ and $f^{-1}([0,4]) = [-2,2]$. Thus

$$f^{-1}([-2,4]) = [-2,2].$$

Remark. We could have also divided the interval $[-2,4]$ into $[-2,0] \cup [0,4]$ or $[-2,0] \cup (0,4]$ (but not $[-2,0) \cup (0,4]!$).

(7) We leave it to readers to check that

$$f^{-1}([-2,1]) = [-1,1].$$

Exercise 2.3.6. Define $f : \mathbb{R}^2 \to \mathbb{R}$ by $f(x,y) = x^2 + y^2$. Find
$$f(\mathbb{R}^+ \times \mathbb{R}), \ f^{-1}(\{-1\}), \ f^{-1}(\{0\}), \ f^{-1}([-2,2)),$$
$$f^{-1}((-\infty,0]) \text{ and } f^{-1}([3,+\infty)).$$

Solution 2.3.6.

(1) Readers may check that

$$f(\mathbb{R}^+ \times \mathbb{R}) = \mathbb{R}^+.$$

(2) Clearly

$$f^{-1}(\{-1\}) = \{(x,y) \in \mathbb{R}^2 : x^2 + y^2 = -1\} = \varnothing.$$

(3) We have

$$f^{-1}(\{0\}) = \{(x,y) \in \mathbb{R}^2 : x^2 + y^2 = 0\} = \{(0,0)\}$$

because $x^2 + y^2 = 0$ gives $x = y = 0$ only since $x^2, y^2 \geq 0$ as x, y are real numbers.

(4) We have

$$f^{-1}([-2,2)) = \{(x,y) \in \mathbb{R}^2 : x^2 + y^2 \in [-2,2)\}$$

so

$$f^{-1}([-2,2)) = \{(x,y) \in \mathbb{R}^2 : 0 \leq x^2 + y^2 < 2\},$$

i.e. it is the open full disk ("open" because we have "strict inequality") of radius $\sqrt{2}$ and center $(0,0)$.

(5) We have

$$f^{-1}((-\infty,0]) = \{(x,y) \in \mathbb{R}^2 : x^2 + y^2 \leq 0\} = \{(0,0)\}$$

because for all $(x,y) \in \mathbb{R}^2: x^2 + y^2 \geq 0$.

(6) We have:

$$f^{-1}([3, +\infty)) = \{(x, y) \in \mathbb{R}^2 : x^2 + y^2 \geq 3\}$$

which is the part of the plane outside the disk of radius $\sqrt{3}$ and center $(0, 0)$ (by taking the circle of the same radius and center as well).

Exercise 2.3.7. Are the following functions injective? Surjective? Bijective?

(1) $f : \mathbb{Z} \to \mathbb{Z}$, $n \mapsto f(n) = 2n$,
(2) $f : \mathbb{Z} \to \mathbb{Z}$, $n \mapsto f(n) = -n$,
(3) $f : \mathbb{R} \to \mathbb{R}$, $x \mapsto f(x) = x^2$,
(4) $f : \mathbb{R} \to \mathbb{R}^+$, $x \mapsto f(x) = x^2$,
(5) $f : \mathbb{R}^+ \to \mathbb{R}$, $x \mapsto f(x) = x^2$,
(6) $f : \mathbb{R}^+ \to \mathbb{R}^+$, $x \mapsto f(x) = x^2$,
(7) $f : \mathbb{C} \to \mathbb{C}$, $x \mapsto f(z) = z^2$.

Solution 2.3.7.

(1) f is injective. Indeed, let $n, m \in \mathbb{Z}$ and take $f(n) = f(m)$, i.e. $2n = 2m$, that is, $n = m$. However, f is not surjective for if it were, we would then have

$$\forall m \in \mathbb{Z}, \ \exists n \in \mathbb{N} : \ f(n) = 2n = m.$$

In particular, for $m = 1 \in \mathbb{Z}$, there would be some $n \in \mathbb{N}$ such that $f(n) = 2n = 1$. But, clearly there is *no* integer n such that $2n = 1$! Thus, f is not surjective, thereby it is not bijective either.

(2) f is injective because

$$\forall n, m \in \mathbb{N} : \ (f(n) = f(m) \Longrightarrow -n = -m \Longrightarrow n = m).$$

It is surjective since

$$\forall m \in \mathbb{Z}, \ \exists n = -m \in \mathbb{Z} \text{ such that } f(n) = f(-m) = -(-m) = m.$$

Accordingly, f is bijective.

(3) The function f is neither injective nor surjective. It is not injective because

$$\exists 1, -1 \in \mathbb{R} : \ f(1) = f(-1) = +1 \text{ and } 1 \neq -1.$$

It is not surjective since

$$\exists (-1) \in \mathbb{R}, \ \forall x \in \mathbb{R} : \ f(x) = x^2 \neq -1.$$

(4) In this case, f is surjective but it is not injective (the same counterexample of Answer 3 works here). Let $y \in \mathbb{R}^+$. Taking $x = \sqrt{y} \in \mathbb{R}$ yields

$$f(x) = f(\sqrt{y}) = (\sqrt{y})^2 = y,$$

i.e. we have shown that f is onto.

(5) f is not surjective but it is injective. The same counterexample of Answer 3 works here to show that f is not onto. Let's show that it is injective. Let $x, x' \in \mathbb{R}^+$ be such that $x^2 = x'^2$. Since $x, x' \geq 0$, taking the positive square root, we deduce directly that $x = x'$.

(6) f is bijective. To show its injectivity, one proceeds exactly as in Answer 5, and to show its surjectivity, one proceeds as in Answer 4.

(7) f is surjective but non-injective, and details are left to interested readers.

Exercise 2.3.8. The same questions as in the previous exercise for the following functions:

(1) $f : \mathbb{N} \to \mathbb{N}$, $n \mapsto f(n) = n + 1$;
(2) $f : \mathbb{Z} \to \mathbb{Z}$, $n \mapsto f(n) = n + 1$;
(3) $f : \mathbb{R}^2 \to \mathbb{R}^2$, $(x, y) \mapsto f(x, y) = (x + y, x - y)$.

Solution 2.3.8.

(1) f is injective but non-surjective (0 does not admit a preimage)...

(2) f is a bijection...

(3) f is bijective. We either show that f is injective and surjective separately, or we show directly that for any $(x', y') \in \mathbb{R}^2$, there is a *unique* couple (x, y) such that

$$f(x, y) = (x + y, x - y) = (x', y').$$

We choose the latter. To this end, let $(x', y') \in \mathbb{R}^2$. By solving the system

$$\begin{cases} x' = x + y, \\ y' = x - y, \end{cases}$$

we see that it has the following unique solution

$$\begin{cases} x = \frac{x'+y'}{2}, \\ y = \frac{x'-y'}{2}. \end{cases}$$

This shows that f is a one-to-one correspondence.

Exercise 2.3.9. Decide, giving reasons, whether each of the following mappings is injective, surjective or bijective:

(1) $f : \mathbb{Z} \times \mathbb{Z} \to \mathbb{Z}$ defined by $f(x, y) = x - y$.

(2) $f : \mathbb{Z} \to \mathbb{Z}$ defined by $f(x) = 3x - 1$.

(3) $f : \mathbb{Q} \to \mathbb{Q}$ defined by $f(x) = 3x - 1$.

(4) $f : \mathbb{R} \to \mathbb{Z}$ defined by $f(x) = [2x]$, where $[\cdot]$ designates the greatest integer function.

Remark. Recall that if $[x]$ denotes the greatest integer function, then for each x, $[x]$ is the largest integer that is less than or equal to x. For example, $[1.2] = 1$, $[0.9] = 0$ and $[-1.4] = -2$.

Solution 2.3.9.

(1) f is not injective because

$$f(0, 1) = f(1, 2) \ (= -1) \text{ but } (0, 1) \neq (1, 2).$$

It is, however, surjective as

$$\forall z \in \mathbb{Z}, \exists (x, y) = (2z, z) \in \mathbb{Z} \times \mathbb{Z} : \ f(2z, z) = 2z - z = z.$$

(2) Clearly, f is injective. But it is not surjective since 0 (for example) does not have a preimage in $\mathbb{Z}$.

(3) Here f is injective and surjective, and so it is bijective.

(4) The function f is not injective because

$$\exists x = \frac{1}{4}, x' = \frac{1}{3} \in \mathbb{R} : \ \left[\frac{2}{4}\right] = \left[\frac{2}{3}\right] = 0 \text{ but } x \neq x'.$$

Finally, f is surjective by the definition of the greatest integer function.

Exercise 2.3.10. Let E be a non-empty set. Show that the mapping $f : \mathcal{P}(E) \to \mathcal{P}(E)$, defined by $f(A) = A^c$ where $\mathcal{P}(E)$ is the powerset of E, is a bijection.

Solution 2.3.10.

(1) Injectivity: We have:

$$\forall A, B \in \mathcal{P}(E) : \ A^c = B^c \implies (A^c)^c = (B^c)^c \implies A = B.$$

(2) Surjectivity: We have:

$$\forall B \in \mathcal{P}(E), \exists A = B^c \in \mathcal{P}(E) : \ f(A) = f(B^c) = (B^c)^c = B.$$

Exercise 2.3.11. Let $f : (0, +\infty) \to \mathbb{R}^+$ and $g : [1, +\infty) \to \mathbb{R}^+$ be two mappings defined by:

$$f(x) = x + \frac{1}{x} \text{ and } g(x) = x + \frac{1}{x},$$

respectively.

(1) Show that f is not injective.
(2) Show that g is injective but not surjective.

Solution 2.3.11.

(1) The function f is not injective because for $2, 1/2 \in (0, +\infty)$, $2 \neq \frac{1}{2}$, one has

$$f(2) = 2 + \frac{1}{2} = \frac{1}{\frac{1}{2}} + \frac{1}{2} = f(1/2).$$

(2) The situation is different here because we cannot take a number and its inverse since we are now working on the set $[1, +\infty)$. To show that g is injective, let $x, x' \geq 1$ be such that $g(x) = g(x')$. Then

$$g(x) = g(x') \iff x + \frac{1}{x} = x' + \frac{1}{x'}$$
$$\iff \frac{x^2 + 1}{x} = \frac{x'^2 + 1}{x'}$$
$$\iff x'x^2 - (x'^2 + 1)x + x' = 0.$$

We need to solve the foregoing equation in x (or in x' if you wish!) by viewing x' as a parameter. We have:

$$\triangle = (x'^2 + 1)^2 - 4 \times x' \times x' = x'^4 + 2x'^2 + 1 - 4x'^2 = (x'^2 - 1)^2 \geq 0,$$

and hence $\sqrt{\triangle} = x'^2 - 1$ for $x' \geq 1$. A priori, there are two "different" solutions given by:

$$x = \frac{x'^2 + 1 - (x'^2 - 1)}{2x'} = \frac{2}{2x'} = \frac{1}{x'}$$

or

$$x = \frac{x'^2 + 1 + x'^2 - 1}{2x'} = x'.$$

We already have $x = x'$, from the second case. If $x = \frac{1}{x'}$, then $xx' = 1$. But $x, x' \geq 1$, that is, $xx' \geq 1$. The only case where we can have $xx' = 1$ under these conditions is $x = x' = 1$. So, in both cases, we obtain $x = x'$. Hence, g is injective.

Finally, g is not surjective since:

$$\exists 1 \in \mathbb{R}^+, \; \forall x \geq 1 : x + \frac{1}{x} = \frac{x^2+1}{x} \neq 1$$

because $x^2 - x + 1 = 0$ has no solution in $\mathbb{R}$.

Exercise 2.3.12. Let f be the map defined from $\mathbb{R}$ into $\mathbb{R}$ by $f(x) = x^3 - x$. Is f injective? Is it surjective?

Solution 2.3.12.

- Injectivity: The function f is not injective because

$$\exists 0, 1 \in \mathbb{R} : \; 1 \neq 0 \text{ and } f(1) = f(0) \; (= 0).$$

- Surjectivity: For the proof, we shall have need for the intermediate value theorem, which is assumed to be known by readers. To show that f is surjective, let y be in $\mathbb{R}$ and we must find at least one real x such that $f(x) = x^3 - x = y$, i.e. we must solve this equation. We know that $x \mapsto \tilde{f}(x) = x^3 - x - y$ is a polynomial (hence continuous) of degree 3. When x is positive and large enough, $\tilde{f}(x)$ is positive, and when x is negative large enough, $f(x)$ is negative. By the intermediate value theorem, $\tilde{f}(x)$ must pass through the real axis at least once. In other words, there is at least one $x \in \mathbb{R}$ such that $f(x) = y$, i.e. f is surjective.

  Remark. There is another way of showing the surjectivity of $f(x) = x^3 - x$. This is usable for any third degree polynomial. It is based upon the Cardano's formulae for solving third degree polynomials. Interested readers are encouraged to investigate this further.

Exercise 2.3.13. Let $f : \mathbb{R} \to \mathbb{R}$ be defined by $f(x) = \sin x$. Is f injective? Is it surjective?

Solution 2.3.13. The sine function is not injective because

$$\exists 0, \pi \in \mathbb{R} : \; f(0) = f(\pi) \; (= 0) \text{ but } 0 \neq \pi.$$

Also, f is not surjective because

$$\exists 2 \in \mathbb{R}, \forall x \in \mathbb{R} : \; f(x) = \sin x \neq 2.$$

Remark. It can directly be shown using basic analysis tools that the sine function defines a bijection from $[-\frac{\pi}{2}, \frac{\pi}{2}]$ onto $[-1, 1]$ (for example).

Exercise 2.3.14. In $\mathbb{R}$, consider $x^4 - x^2 + 1 = \alpha$, where $\alpha \in \mathbb{R}$.

(1) Solve the equation in the special cases $\alpha = -1$ and $\alpha = 1$.

(2) Let f be defined from $\mathbb{R}$ into $\mathbb{R}$ by $f(x) = x^4 - x^2 + 1$. Is f injective? Is it surjective?

Solution 2.3.14.

(1) (a) The case $\alpha = -1$: By setting $y = x^2$ or else, we see that the equation $x^4 - x^2 + 2 = 0$ has no real solutions.

(b) The case $\alpha = 1$: The given equation becomes $x^4 - x^2 = 0$. Its solutions therefore are $x = 0$ or $x = 1$ or $x = -1$.

(2) According to the previous answer, we have $f(0) = f(1)$ which amply implies that f is not injective. By the previous answer once again, $x^4 - x^2 + 1 = -1$ does not have any real solution x, and so

$$\exists y = -1, \forall x \in \mathbb{R} : \ f(x) = x^4 - x^2 + 1 \neq -1,$$

i.e. f is not surjective.

Exercise 2.3.15.

(1) Show that a strictly monotone real function is injective.

(2) Infer that, e.g. $x \mapsto x^n$, defined from $\mathbb{R}$ into $\mathbb{R}$, where n is an odd number, is injective.

Solution 2.3.15.

(1) First, recall that a strictly monotone (or monotonic) function is a function $f : A \to \mathbb{R}$, where A is a subset of $\mathbb{R}$, that is either strictly increasing, i.e. for all $x, y \in A$:

$$x < y \Longrightarrow f(x) < f(y),$$

or strictly decreasing, i.e.

$$x < y \Longrightarrow f(x) > f(y).$$

WLOG, assume that f is strictly increasing, and let $x, y \in A$ be such that $x \neq y$. Then either $x < y$ or $x > y$.

(a) If $x < y$, given that f is strictly increasing, it must follow that $f(x) < f(y)$, and so $f(x) \neq f(y)$.

(b) If $x > y$, then it ensues that $f(x) > f(y)$ again as f is strictly increasing, from which we derive yet again that $f(x) \neq f(y)$.

Thus, and in either case, $f(x) \neq f(y)$. In other words, we have shown that f is one-to-one.

Remark. Once we are in, we must say something about the converse. Even though it might be perceived as extraneous to the present manuscript, it is primordial for readers to know. So, does an injective function have to be strictly monotone? The answer is negative in general. For example, the function $f : \mathbb{R}^* \to \mathbb{R}$ defined by $f(x) = 1/x$ is injective (and continuous) but it is not monotone (do the details). Also, the function $f : \mathbb{R} \to \mathbb{R}$ (observe that f is defined on an interval this time) defined by

$$f(x) = \begin{cases} x, & x \in \mathbb{R} \setminus \{1, -1\}, \\ -1, & x = 1, \\ 1, & x = -1, \end{cases}$$

is one-to-one (and non-continuous), but f is not strictly monotone. Nonetheless, a *continuous* one-to-one function which is defined on an *interval* is strictly monotone (the proof is based upon the intermediate value theorem).

(2) The function $f : \mathbb{R} \to \mathbb{R}$ defined by $f(x) = x^n$, where is odd, is strictly increasing. To see that, we use a proof by induction. Let $x, y \in \mathbb{R}$ be such that $x < y$.

   (a) The case $0 \le x < y$: It is clear that $x^1 < y^1$. Assume that $x^n < y^n$, and let us show that $x^{n+1} < y^{n+1}$. By the induction hypothesis, and $0 < x < y$, we may write

   $$x^{n+1} = x^n x < y^n x < y^n y = y^{n+1}.$$

   (b) The case $x \le 0 < y$: In this case, $x^n \le 0$ as the n is odd, and $y^n > 0$. So, $x^n < y^n$.

   (c) The case $x < y \le 0$: Clearly, $0 \le -y < -x$. By the first case, $(-y)^n < (-x)^n$ or $-y^n < -x^n$ or $x^n < y^n$, as needed.

   Remark. The function $f : \mathbb{R} \to \mathbb{R}$, defined by $f(x) = x^n$, is not injective for any even n, as, e.g. $1 \ne -1$ and yet

   $$f(1) = 1^n = 1 = (-1)^n = f(-1)$$

   for every even natural number n.

Exercise 2.3.16. ([9]) What are the real solutions (x, y) of the equation

$$x^4 + x^3 y + x^2 y^2 + x y^3 + y^4 = 0?$$

Solution 2.3.16. First, it is seen that $(x, y) = (0, 0)$ is an evident solution of the equation $x^4 + x^3 y + x^2 y^2 + x y^3 + y^4 = 0$. Is there any other solution? We will show that there isn't any! If $x = y$, the previous equation gives $5x^4 = 0$, whereby $x = 0$ (and so $y = 0!$). So, assume $x \neq y$. Since $x \mapsto x^5$, defined from $\mathbb{R}$ into $\mathbb{R}$, is injective by Exercise 2.3.15, we have $x^5 \neq y^5$ or $x^5 - y^5 \neq 0$. However,

$$x^5 - y^5 = (x - y)(x^4 + x^3 y + x^2 y^2 + x y^3 + y^4) = 0.$$

This contradiction, combined with the above arguments, signify that the only solution of the above equation if $(x, y) = (0, 0)$.

Exercise 2.3.17. It is known from a basic chemistry course say that the formula which converts Celsius temperature to Fahrenheit temperature is given by $f(x) = 9/5x + 32$. This defines a bijection from $\mathbb{R}$ onto $\mathbb{R}$. What is the formula that converts Fahrenheit temperature to Celsius temperature?

Solution 2.3.17. This is a simple application of inverse functions. The sought formula is the inverse function f^{-1}. To find it, set $y = f(x)$, then we write x in terms of y. We have

$$y = 9/5x + 32 \iff y - 32 = 9/5x \iff x = 5/9(y - 32).$$

Therefore, the required formula that converts Fahrenheit temperature to Celsius temperature is given by

$$f^{-1}(x) = 5/9(x - 32).$$

Exercise 2.3.18. Let $f : E \to F$ $(E, F \subset \mathbb{R})$ be a function defined by:

$$f(x) = \frac{x + 1}{2x - 3}.$$

Provide the largest E and F so that f becomes bijective.

Solution 2.3.18. First, f must be well-defined, and so E must not contain the point $\frac{3}{2}$. Take for example, $E = \mathbb{R} \setminus \{\frac{3}{2}\}$. Let's find F such that f is surjective. Each y must have a preimage in $E = \mathbb{R} \setminus \{\frac{3}{2}\}$. Let us write x in terms of y. We have:

$$y = \frac{x + 1}{2x - 3} \iff (2x - 3)y = x + 1$$
$$\iff 2xy - 3y = x + 1$$
$$\iff 2xy - x = 3y + 1$$
$$\iff x(2y - 1) = 3y + 1$$
$$\iff x = \frac{3y + 1}{2y - 1}.$$

One can clearly see that $\frac{1}{2}$ does not have a preimage. We also see that for each $y \neq 1/2$, there exists a unique $x \neq 3/2$ such that $f(x) = y$. We therefore infer that f is bijective from $E = \mathbb{R} \setminus \{\frac{3}{2}\}$ onto $F = \mathbb{R} \setminus \{\frac{1}{2}\}$.

Exercise 2.3.19.

(1) Find ran f in the following cases:
    (a) $f : \mathbb{N} \to \mathbb{N}$ defined by $f(x) = 2x + 1$.
    (b) $f : \mathbb{Z} \to \mathbb{Q}$ defined by $f(x) = \frac{x}{2}$.
    (c) $f : \mathbb{Z} \times \mathbb{Z} \to \mathbb{Z}$ defined by $f(x, y) = y - x$.
(2) Deduce, among these functions, those that are surjective.

Solution 2.3.19.

(1) (a) We have
$$\operatorname{ran} f = \{2x + 1 : x \in \mathbb{N}\} = \{1, 3, 5, 7, \cdots\}.$$

(b) We have
$$\operatorname{ran} f = \left\{\frac{x}{2} : x \in \mathbb{Z}\right\} = \left\{0, \pm\frac{1}{2}, \pm\frac{2}{2}, \pm\frac{3}{2}, \cdots\right\}.$$

(c) We find
$$\operatorname{ran} f = \mathbb{Z}.$$

(2) Only the third function is surjective.

Exercise 2.3.20. Investigate the injectivity and the surjectivity of the function $f : \mathbb{R}^2 \to \mathbb{R}^2$ defined by
$$f(x, y) = (x + y, xy).$$

Solution 2.3.20.

(1) f is not injective because
$$\exists (1, 2), (2, 1) \in \mathbb{R}^2, \ (1, 2) \neq (2, 1) \text{ but } f(1, 2) = f(2, 1) \ (= (3, 2)).$$

(2) f is not surjective since:
$$\exists (0, 1) \in \mathbb{R}^2, \forall (x, y) \in \mathbb{R}^2 : (0, 1) \neq f(x, y) = (x + y, xy)$$

because the system $\begin{cases} x + y = 0 \\ xy = 1 \end{cases}$ has no solution in $\mathbb{R}^2$, as may be checked.

Exercise 2.3.21. Let $f : \mathbb{N} \times \mathbb{N} \to \mathbb{N}$ be the mapping defined by:
$$f(n, m) = 2^n 3^m.$$
Show that f injective.

Solution 2.3.21.

(1) Let $(n, m), (n', m') \in \mathbb{N}^2$ be such that $f(n, m) = f(n', m')$, i.e.
$$2^n 3^m = 2^{n'} 3^{m'}.$$

If $n \neq n'$, then either $n < n'$ or $n > n'$. If $n < n'$, then

$$2^n 3^m = 2^{n'} 3^{m'} \implies \underbrace{3^m}_{\text{odd}} = \underbrace{2^{n'-n} 3^{m'}}_{\text{even}},$$

which is absurd!

If $n > n'$, then

$$2^n 3^m = 2^{n'} 3^{m'} \implies \underbrace{2^{n-n'} 3^m}_{\text{even}} = \underbrace{3^{m'}}_{\text{odd}},$$

which is again absurd! So, necessarily $n = n'$. Now, we show that $m = m'$. Since $n = n'$,

$$2^n 3^m = 2^{n'} 3^{m'} \implies 3^m = 3^{m'} \implies \log_3 3^m = \log_3 3^{m'} \implies m = m'.$$

Hence f is injective, as suggested.

Exercise 2.3.22. Let A be the set of odd integers, then define the function $f : \mathbb{N} \to A$ as follows:

$$f(x) = \begin{cases} x, & \text{if } x \text{ is odd,} \\ 1 - x, & \text{if } x \text{ is even.} \end{cases}$$

Show that f bijective.

Solution 2.3.22. First, observe that:

$$f(x) > 0 \text{ if } x \text{ is odd, and } f(x) < 0 \text{ if } x \text{ is even.}$$

First, we show that f is one-to-one. Let $x, y \in \mathbb{N}$ be such that $f(x) = f(y)$. Clearly, either $f(x)$ and $f(y)$ are both positive or both negative, otherwise $f(x) \neq f(y)$.

(1) If $f(x)$ and $f(y)$ are positive at the same time, and if $f(x) = f(y)$, then since $f(t) = t$ in this case, we get $x = y$.

(2) If $f(x)$ and $f(y)$ are both negative, and if $f(x) = f(y)$, then we have in this case, $1 - x = 1 - y$ or merely $x = y$.

Now, we show that f is surjective. Let $y \in A$ (then $y \neq 0$). We must find at least one x for which $y = f(x)$.

(1) If $y > 0$, then take $x = y$ and so $f(x) = f(y) = y$ since x is odd.

(2) If $y < 0$, by taking $x = 1 - y$, we see that $1 - y$ becomes even (and positive). We then get

$$f(x) = f(1 - y) = 1 - (1 - y) = y,$$

and the proof is complete.

Exercise 2.3.23. Let $f : \mathbb{R} \to \mathbb{R}$ be defined by $f(x) = x^2 - 3x + 2$. f is injective? Surjective?

Solution 2.3.23. By an appropriate change of variables with respect to the Cartesian coordinate system, the function f may be viewed as say $F : \mathbb{R} \to \mathbb{R}$ with $F(X) = X^2$. Since F is neither injective nor surjective, it is expected that f behaves like F. Let us corroborate this by rigorous proofs.

The function f is not injective because

$$\exists x = 1, x' = 2 \in \mathbb{R} : \ x \neq x' \text{ and } f(x) = f(x') \ (= 0).$$

The function is not surjective either. If we want to find the right interval of counterexamples, we will have to find out what goes wrong if we had to "show" that f is surjective. Let $y \in \mathbb{R}$, then we try to find a certain $x \in \mathbb{R}$ such that $x^2 - 3x + 2 = y$ or $x^2 - 3x + 2 - y = 0$. The discriminant $\triangle$ is given by:

$$\triangle = 9 - 4(2 - y) = 1 + 4y.$$

If $1 + 4y \geq 0$, the equation admits solutions, and if $1 + 4y < 0$, the equation does not admit any solution. Since $1 + 4y$ is not positive for every $y \in \mathbb{R}$, f is not surjective. More precisely, e.g.

$$\exists y = -1 \in \mathbb{R} : \ \forall x \in \mathbb{R} : \ x^2 - 3x + 2 \neq -1$$

because $x^2 - 3x + 3$ never vanishes! Thus f is not onto.

Remark. Do not make the mistake of saying that for y such that $1 + 4y \geq 0$, f is surjective, and for y such that $1 + 4y > 0$ f is not surjective, as this is non-sense!

What is correct, for instance, is to say that $g : \mathbb{R} \to [-1/4, \infty)$ (not f!), defined by $g(x) = x^2 - 3x + 2$, is surjective.

Exercise 2.3.24. Let E and F be two finite sets, and let $f : E \to F$ be a function. Show the following statements:

(1) If f is injective, then $\operatorname{card} E \leq \operatorname{card} F$.
(2) If f is surjective, then $\operatorname{card} E \geq \operatorname{card} F$.
(3) Infer that if f is bijective, then $\operatorname{card} E = \operatorname{card} F$.
(4) Let $\operatorname{card} E = \operatorname{card} F$. Show that if f is injective, then f is surjective. Likewise, if f is surjective, then f is injective. Accordingly, if $E = F$, and so $f : E \to E$, then

$$f \text{ injective} \iff f \text{ surjective} \iff f \text{ bijective}.$$

Solution 2.3.24.

(1) As f is injective, each element of E is sent to a different element of F. This says that we must have $\operatorname{card} E \leq \operatorname{card} F$.
(2) To reach a contradiction, assume $\operatorname{card} E < \operatorname{card} F$. To each element of E, there corresponds a unique element of F. So, it

becomes evident that some element (s) is (are) not the image of any $x \in E$. Hence $f(E) \subsetneq F$, and this contradicts the assumption that f is surjective. Thus, card $E \geq$ card F.

(3) This is clear by the two preceding questions.

(4) Suppose that f is one-to-one. Then $f : E \to f(E)$ is a bijection (with a flagrant abuse of notation!). Hence card $E =$ card $f(E)$. By assumption card $E =$ card F, and so card $f(E)$ $=$ card F. Since $f(E) \subset F$ all the time, we deduce that $f(E) = F$, thereby showing the ontoness of f.

Conversely, suppose f is surjective. If f were not injective, then there would be at least two different elements of E, noted a and b, such that $f(a) = f(b)$. This would entail card $f(E) <$ card E. Since by hypothesis card $E =$ card F, it would ensue that card $f(E) <$ card F. But hang on, this would contradict the surjectivity of f! Thus, f must be injective.

To conclude, observe that when $f : E \to E$, then f is injective if and only if it is surjective. Therefore, if f is injective, it is surjective and so it becomes bijective, and the circle is complete.

Remark. Recall that a bijection from E onto E, where E is a finite set, is called a permutation.

Exercise 2.3.25. (Cf. Exercise 2.4.11) Let E and F be two finite sets (for example, card $E = n$ and card $F = m$).

(1) What is the number of functions from E into F?
(2) What is the number of one-to-one functions from E into F?
(3) What is the number of bijective functions from E onto F, where card $E =$ card $F = n$?

Solution 2.3.25.

(1) Let $f : E \to F$ be a function. Each x in E has m choices for $f(x)$. So, it is pretty clear that the number of functions from E into F is m^n.

(2) Assume that the elements of E are $x_1, x_2, \cdots, x_n$, and those of F are $y_1, y_2, \cdots, y_m$. Then x_1 could be assigned to any of the $y_i, i = 1, \cdots, m$, that is, we have m choices. Since f is one-to-one, there are now only $(m-1)$ choices left for x_2, etc. Consequently, the number of injections from E into F is given by:
$$m(m-1)\cdots(m-(n-1))$$
which is nothing but $m!/(m-n)!$.

(3) This is now easy to see. In virtue of Exercise 2.3.24, f is injective if and only if it is bijective. Since card $E =$ card $F = n$, the foregoing question tells us that the number of bijections is

$$n(n-1)\cdots(n-(n-1)) = n \times (n-1) \times \cdots \times 2 \times 1 = n!.$$

Exercise 2.3.26. Let E, F and G be three non-empty sets. Let $f : E \to F$ and $g : F \to G$ be two functions. Show that:

(1) If f and g are injective, then $g \circ f$ is injective.
(2) If f and g are surjective, then $g \circ f$ is surjective.

Solution 2.3.26. Recall that $g \circ f : E \to G$.

(1) Assume that f and g are injective. To prove that $g \circ f$ is one-to-one, let $x, y \in E$ be such that $(g \circ f)(x) = (g \circ f)(y)$. Since g is injective, we have $f(x) = f(y)$. Since f is injective, we get $x = y$. Thus, $g \circ f$ is injective.

(2) Suppose f and g are surjective. To show that $g \circ f$ is surjective, let $z \in G$. Since g is surjective, there exists at least a y in F such that $g(y) = z$. Since $y \in F$ and f is surjective, there is at least an x in E such that $f(x) = y$. So we have shown that for all $z \in G$, there exists at least an $x \in E$ such that

$$(g \circ f)(x) = g(f(x)) = g(y) = z,$$

which means that $g \circ f$ is surjective.

Remark. We may also show the previous result as follows: Since f and g are surjective, $f(E) = F$ and $g(F) = G$. To show that $g \circ f$ is surjective, we show that $(g \circ f)(E) = G$. Clearly

$$(g \circ f)(E) = g(f(E)) = g(F) = G,$$

as wished.

Exercise 2.3.27. Let E, F and G be three non-empty sets. Let $f : E \to F$ and $g : F \to G$ be two functions. Prove that:

(1) If $g \circ f$ is injective, then f is injective.
(2) If $g \circ f$ is injective and f surjective, then g is injective.
(3) If $g \circ f$ is surjective, then g is surjective.
(4) If $g \circ f$ is surjective and g is injective, then f is surjective.

Solution 2.3.27. Remember that $g \circ f$ is defined from E into G.

(1) Assume that $g \circ f$ is injective, and let $x, x' \in E$ be such that $f(x) = f(x')$. Hence

$$g(f(x)) = g(f(x')) \text{ or } (g \circ f)(x) = (g \circ f)(x').$$

But $g \circ f$ is injective, and so $x = x'$, i.e. we have shown that f is injective.

(2) Suppose $g \circ f$ is injective and f is surjective. Let $y, y' \in F$ be such that $g(y) = g(y')$. Since f is surjective, we know that

$$\exists x \in E : \ y = f(x) \text{ and } \exists x' \in E : y' = f(x').$$

Whence

$$g(f(x)) = g(y) = g(y') = g(f(x')).$$

But $g \circ f$ is injective, and so $x = x'$, that is, $f(x) = f(x')$ or $y = y'$. Ergo, g is injective.

(3) Suppose that $g \circ f$ is surjective. To show that g is surjective, let $z \in G$. Since $g \circ f$ is surjective,

$$\exists x \in E, \ (g \circ f)(x) = g(f(x)) = z.$$

Take $y = f(x)$. It is an element of F which satisfies $g(y) = z$. In other words, we have shown that g is surjective.

(4) Assume that $g \circ f$ is surjective and that g is injective. Let $y \in F$, and so $g(y) \in G$. Since $g \circ f$ is surjective, there exists (at least) an $x \in E$ such that $g(y) = g \circ f(x) = g(f(x))$. By the injectivity of g, we get $y = f(x)$, i.e. f is surjective, as needed.

Remark. By the first and third questions of the previous exercise, we easily deduce that:

$$g \circ f \text{ bijective} \implies f \text{ injective and } g \text{ surjective}.$$

Exercise 2.3.28. Define $\mathbb{Z}[\sqrt{2}] = \{a + b\sqrt{2} : a, b \in \mathbb{Z}\}$ and $\mathbb{Z}[\sqrt{3}] = \{a + b\sqrt{3} : a, b \in \mathbb{Z}\}$. Let $f : \mathbb{Z}[\sqrt{2}] \to \mathbb{Z}[\sqrt{3}]$ be a function defined by

$$f(a + b\sqrt{2}) = a + b\sqrt{3}.$$

Show that f is injective.

Solution 2.3.28. Let $x, x' \in \mathbb{Z}[\sqrt{2}]$ and assume $f(x) = f(x')$. By definition, $x = a + b\sqrt{2}$ and $x' = a' + b'\sqrt{2}$, with $a, b; a', b' \in \mathbb{Z}$. We ought to show that $x = x'$, i.e. $a = a'$ and $b = b'$. We have

$$f(x) = f(x') \implies a + b\sqrt{3} = a' + b'\sqrt{3}$$
$$\implies a - a' = (b' - b)\sqrt{3}.$$

If $b \neq b'$, then

$$\sqrt{3} = \frac{a - a'}{b' - b},$$

which is a contradiction because $\sqrt{3} \notin \mathbb{Q}$ (see Exercise 1.3.27) whilst $\frac{a-a'}{b'-b} \in \mathbb{Q}$ (because $a - a' \in \mathbb{Z}$ and $b' - b \in \mathbb{Z}^*$). Therefore, we must have $b = b'$. But this leads to

$$a - a' = (b' - b)\sqrt{3} = 0 \times \sqrt{3} = 0, \text{ i.e. } a = a'.$$

Thus, $x = x'$, i.e. f is injective.

Exercise 2.3.29. Find the inverse of each of the following bijective functions:

(1) $f : \mathbb{R}^+ \to \mathbb{R}^+$, $x \mapsto f(x) = x^2$;
(2) $f : \mathbb{Q} \to \mathbb{Q}$, $x \mapsto 3x - 1$;
(3) $f : \mathbb{R}^2 \to \mathbb{R}^2$, $(x, y) \mapsto f(x, y) = (x + y, x - y)$;
(4) $f : \mathcal{P}(E) \to \mathcal{P}(E)$, $A \mapsto f(A) = A^c$, where E is a non-void set.

Solution 2.3.29.

(1) To find the inverse function, we write x in terms of y from $y = f(x) = x^2$. Since x and y are both positive, we can write $x = \sqrt{y}$. The inverse function is therefore given by:

$$f^{-1} : \mathbb{R}^+ \to \mathbb{R}^+, \ x \mapsto f^{-1}(x) = \sqrt{x}.$$

(2) Let $y \in \mathbb{Q}$ be such that $y = f(x)$. We have

$$y = 3x - 1 \iff y + 1 = 3x \iff x = \frac{1}{3}(y + 1).$$

So, the inverse function is defined by:

$$f^{-1} : \mathbb{Q} \to \mathbb{Q}, \ x \mapsto f^{-1}(x) = \frac{1}{3}(x + 1).$$

(3) Let $(x', y') \in \mathbb{R}^2$ be such that $f(x, y) = (x + y, x - y) = (x', y')$. By solving the system

$$\begin{cases} x' = x + y, \\ y' = x - y, \end{cases}$$

we see that the unique solution is given by

$$\begin{cases} x = \frac{x' + y'}{2}, \\ y = \frac{x' - y'}{2}. \end{cases}$$

Hence,

$$f^{-1} : \mathbb{R}^2 \to \mathbb{R}^2, \ (x, y) \mapsto f^{-1}(x, y) = \left(\frac{1}{2}(x + y), \frac{1}{2}(x - y) \right).$$

(4) It is clear that the inverse function is given by:

$$f^{-1} : \mathcal{P}(E) \to \mathcal{P}(E), \ A \mapsto f^{-1}(A) = A^c,$$

i.e. the function, this time, is equal to its inverse (such a function is often called an involution).

Exercise 2.3.30. Prove that the function $f : \mathbb{R} \to \mathbb{R}$ defined for all $x \in \mathbb{R}$ by $f(x) = 0$ is not invertible by using two methods.

Solution 2.3.30.

(1) First method: We apply the definition. If f *were* invertible, then there would be a function $g : \mathbb{R} \to \mathbb{R}$ such that:

$$f \circ g = g \circ f = \mathrm{id}_{\mathbb{R}}, \ \text{ i.e. } f(g(x)) = g(f(x)) = x, \ \forall x \in \mathbb{R}.$$

But, $f(g(x)) = 0$ for each x, and so we would not have $f(g(x)) = x$ for all real x!

(2) Second method: We know that "invertible" and "bijective" are two equivalent notions in this context. However, we can see clearly that f is not bijective because it is not injective for $f(0) = f(1) \ (= 0)$ yet $0 \neq 1$.

Exercise 2.3.31.

(1) Let $f : E \to E$ be a function such that $f \circ f = \mathrm{id}_E$. Show that f is bijective.

(2) Deduce that the map $f : [0, 1] \to [0, 1]$ defined by:

$$f(x) = \begin{cases} x, & \text{if } x \in [0, 1] \cap \mathbb{Q}, \\ 1 - x, & \text{otherwise}, \end{cases}$$

is bijective.

Solution 2.3.31.

(1) We can answer this question using different methods. For example, since $f \circ f = \mathrm{id}_E$, we have $f \circ f = f \circ f = \mathrm{id}_E$. So f is invertible and $f^{-1} = f$.

(2) According to the previous question, to prove that f is bijective, it suffices to check that $f \circ f = \mathrm{id}_{[0,1]}$. If $x \in [0, 1] \cap \mathbb{Q}$, then $f(x) = x \in [0, 1] \cap \mathbb{Q}$, and so

$$(f \circ f)(x) = f(f(x)) = f(x) = x.$$

If $x \notin [0, 1] \cap \mathbb{Q}$, then $f(x) = 1 - x$. Hence

$$(f \circ f)(x) = f(1 - x).$$

But $(1 - x) \notin [0, 1] \cap \mathbb{Q}$, and so $f(1 - x) = 1 - (1 - x) = x$. Thus

$$\forall x \in [0, 1]: \ f \circ f(x) = x = \mathrm{id}_{[0,1]}(x),$$

as needed.

Exercise 2.3.32. Investigate the invertibility of the following functions, and give the inverse when the latter exists.

(1) $f : \mathbb{R} - \{2\} \to \mathbb{R} - \{2\}$, $x \mapsto f(x) = \frac{2x-1}{x-2}$;

(2) $f : \mathbb{Z}^2 \to \mathbb{Z}$, $(n, m) \mapsto f(n, m) = nm^2$;

(3) $f : \mathbb{R}^2 \to \mathbb{R}^2$, $(x, y) \mapsto f(x, y) = (y, -x)$;

(4) $f : [\frac{3}{2}, \infty) \to [-\frac{1}{4}, \infty)$, $x \mapsto f(x) = x^2 - 3x + 2$.

Solution 2.3.32.

(1) The function f is bijective (we have already treated a similar example), i.e. f is invertible. To find its inverse, we write x in terms of y ($= f(x)$). For all $x, y \neq 2$, we have

$$y = \frac{2x - 1}{x - 2} \iff y(x - 2) = 2x - 1$$

$$\iff xy - 2y = 2x - 1$$

$$\iff xy - 2x = 2y - 1$$

$$\iff x(y - 2) = 2y - 1$$

$$\iff x = \frac{2y - 1}{y - 2}.$$

Hence, the inverse is given by

$$f^{-1} : \mathbb{R} - \{2\} \to \mathbb{R} - \{2\}, \ x \mapsto f^{-1}(x) = \frac{2x - 1}{x - 2} \ (= f(x))$$

(it is an involution).

(2) The function f is not injective. For instance, witness $(1, -1)$ and $(1, 1)$. Therefore, f cannot be invertible.

(3) To obtain the inverse of f (which also constitutes a proof of the bijectivity of f), let $(x, y) \in \mathbb{R}^2$ and $(x', y') \in \mathbb{R}^2$. We get

$$(x', y') = f(x, y) = (y, -x) \iff x = -y' \text{ and } y = x'.$$

The inverse function is therefore given by

$$f^{-1} : \mathbb{R}^2 \to \mathbb{R}^2, (x, y) \mapsto f^{-1}(x, y) = (-y, x).$$

(4) Let's prove that f is bijective by showing that for all $y \geq -\frac{1}{4}$ there is a *unique* $x \geq \frac{3}{2}$ such that $y = f(x)$. Let $y \geq -\frac{1}{4}$ be such that $y = f(x) = x^2 - 3x + 2$ or $x^2 - 3x + 2 - y = 0$. The discriminant is given by $\Delta = 1 + 4y$. So, there are two solutions of this equation given by:

$$x = \frac{3 - \sqrt{1 + 4y}}{2} \text{ or } x = \frac{3 + \sqrt{1 + 4y}}{2}.$$

Since $x \geq \frac{3}{2}$, we may only accept the second solution (we may take the first one only when $y = -\frac{1}{4}$, but in this case the two solutions would be equal!). Accordingly, there exists always a unique solution, given by:

$$x = \frac{3 + \sqrt{1 + 4y}}{2}.$$

The inverse function is therefore given by

$$f^{-1} : [-1/4, \infty) \to [3/2, \infty), \ x \mapsto f^{-1}(x) = \frac{3 + \sqrt{1 + 4x}}{2}.$$

2.4. Supplementary Exercises

Exercise 2.4.1. Let $f : \mathbb{R} \to \mathbb{R}$ be the function defined by

$$f(x) = |x|, \ \forall x \in \mathbb{R}.$$

Find

$$f(\{-2, 1\}), \ f([-2, -1]), \ f([-2, 1]), \ f^{-1}(\{2\}) \text{ and } f^{-1}([-2, 3]).$$

Exercise 2.4.2. Let $f : \mathbb{R} \to \mathbb{R}$ be a mapping defined by

$$f(x) = \sin(\pi x).$$

Find:

$$f(\{0\}), \ f(\mathbb{Z}), \ f([0, 1/2]), \ f^{-1}(\{2\}) \text{ and } f^{-1}(\{1\}).$$

Exercise 2.4.3. Let f be the map defined from $\mathbb{R}$ into $\mathbb{R}$ by $f(x) = x^3 + x$. Is f injective? Is it surjective?

Exercise 2.4.4. Let $f : \mathbb{R} \setminus \{1/2\} \to \mathbb{R}$ be a map defined by

$$f(x) = \frac{x + 1}{2x - 1}.$$

Prove that f is injective. Is f bijective?

Exercise 2.4.5. Let $f : \mathbb{R} \to \mathbb{R}$ be defined by

$$f(x) = \frac{2x}{x^2 + 1}.$$

(1) Investigate the injectivity and surjectivity of f.
(2) Show that $f(\mathbb{R}) = [-1, 1]$.
(3) Show that the restriction of f, denoted by g, which is defined from $[-1, 1]$ into $[-1, 1]$ by $g(x) = \frac{2x}{x^2+1}$, is bijective.

Exercise 2.4.6. Let $f : \mathbb{R} \to \mathbb{R}$ be a map defined by

$$f(x) = -x^5 + x^4 + x^3 - 1.$$

Is f injective? Is it surjective?

Exercise 2.4.7. Let $f : \mathbb{R}^2 \to \mathbb{R}$ and $g : \mathbb{R} \to \mathbb{R}^2$ be defined by:

$$f(x, y) = xy \text{ and } g(x) = (x, x^2),$$

respectively.

(1) Decide whether f and g is one-to-one, onto or bijective.
(2) Find $f \circ g$ and $g \circ f$.
(3) Is $f \circ g$ injective? Is it surjective? Is it bijective? The same questions for $g \circ f$.

Exercise 2.4.8. Let E be a set. Let $f : E \to E$ be a map such that $f^3 = f$ where $f^3 = f \circ f \circ f$. Prove that we have the equivalence among the following statements:

(1) f injective;
(2) f surjective;
(3) f bijective.

Exercise 2.4.9. Let E be a non-empty set. Let $f, g, h : E \to E$ be three mappings. Show that:

$$f \text{ is injective} \iff (\forall g, h : f \circ g = f \circ h \Rightarrow g = h).$$

Hint: For the implication "$\Leftarrow$", take $h = \mathrm{id}_E$ on E, and g the function defined by:

$$g(x) = \begin{cases} y, & x \in E \setminus \{x\}, \\ x', & y = x, \end{cases}$$

where $y \in E$.

Exercise 2.4.10. Let α and β be two real numbers. Let the function $\psi_{\alpha,\beta} : \mathbb{R} \to \mathbb{R}$ be defined by:

$$\psi_{\alpha,\beta}(x) = \alpha x + \beta.$$

(1) Find all the values of α and β for which $\psi_{\alpha,\beta}$ is bijective.
(2) In the case where $\psi_{\alpha,\beta}$ is bijective find its inverse.

Exercise 2.4.11. Let E and F be two finite sets, such that card $E = m$ and card $F = n$. Show that the number of surjective functions from E onto F is given by:

$$n^m + \sum_{k=1}^{n-1} (-1)^k \binom{n}{k} (n-k)^m.$$

Exercise 2.4.12. (Cf. exercise 2.3.10) Let $f : E \to F$ be a function. Show that

$$f \text{ is bijective} \iff \forall A \in \mathcal{P}(E) : f(A^c) = (f(A))^c,$$

where $\mathcal{P}(E)$ is the powerset of E.

CHAPTER 3

Binary Relations

3.1. Basics

DEFINITION 3.1.1. A binary relation $\mathcal{R}$ on a set X is a subset $A_{\mathcal{R}}$ of $X \times X$. As it is customary, instead of writing $(x, y) \in A_{\mathcal{R}}$, we will usually write $x\mathcal{R}y$.

DEFINITION 3.1.2. Let $\mathcal{R}$ be a binary relation on X. We say that $\mathcal{R}$ is:

(1) reflexive if: $\forall x \in X$: $x\mathcal{R}x$.
(2) symmetric if: $\forall x, y \in X$: $x\mathcal{R}y \Rightarrow y\mathcal{R}x$.
(3) transitive if: $\forall x, y, z \in X$: $(x\mathcal{R}y$ and $y\mathcal{R}z) \Rightarrow x\mathcal{R}z$.

When $\mathcal{R}$ is reflexive, symmetric and transitive, then $\mathcal{R}$ is called an equivalence relation on X.

DEFINITION 3.1.3. Let $\mathcal{R}$ be an equivalence relation on X. The equivalent class of $x \in X$, denoted by $\dot{x}$ (or $\overline{x}$), is defined by:

$$\dot{x} = \{y \in X : y\mathcal{R}x\}.$$

The collection of all equivalent classes of X, denoted by $X/\mathcal{R}$ is called the quotient of X by $\mathcal{R}$, that is,

$$X/\mathcal{R} = \{\dot{x} : x \in X\}.$$

Remark. Equivalent relations are concerned with the notion of equality in a certain sense.

Now, we give basic properties of the quotient.

PROPOSITION 3.1.1. *Let $\mathcal{R}$ be an equivalence relation on X and let $\dot{x}$ be the equivalent class of $x \in X$. Then:*

(1) $\forall x \in X$: $x \in \dot{x}$.
(2) $\dot{x} = \dot{y}$ iff $x\mathcal{R}y$.
(3) If $\dot{x} \neq \dot{y}$, then $\dot{x}$ and $\dot{y}$ must be disjoint.

After having introduced the equivalence relation, we define an equally important class of binary relations.

DEFINITION 3.1.4. Let $\mathcal{R}$ be a relation on X. If $\mathcal{R}$ is reflexive, anti-symmetric and transitive, then we say that $\mathcal{R}$ is an order relation on X.

Remark. Order relations are concerned with ranking elements of some set with respect to that relation.

DEFINITION 3.1.5. Let $\mathcal{R}$ be an order relation on X. If all elements of X are comparable, i.e.

$$\forall x, y \in X : x\mathcal{R}y \text{ or } y\mathcal{R}x,$$

then $\mathcal{R}$ is said a total order. Otherwise, we say that R is a partial order.

We finish by giving a generalization of the notions "least upper/ greatest lower bounds", already known in $\mathbb{R}$ with respect to the order relation "$\leq$".

DEFINITION 3.1.6. Let E be a non-empty set and $A \subset E$, and let $\mathcal{R}$ be an order relation over E. The element $b \in A$ is the largest element of A if for all $x \in A$: $x\mathcal{R}b$. In such case, b is denoted by $\max A$.

Likewise, call $a \in A$ the smallest element of A if for all $x \in A$: $a\mathcal{R}x$. If that is the case, a is noted $\min A$.

We say that $a \in E$ (resp. $b \in E$) is a lower bound (resp. an upper bound) for A if

$$\forall x \in A : a\mathcal{R}x \text{ (resp. } x\mathcal{R}b).$$

If the set of all lower bounds for A has a largest element, it is called the greatest lower bound or infimum. It is noted $\inf A$.

If the set of all upper bounds for A has a smallest element, it is called the least upper bound or supremum. It is denoted by $\sup A$.

Remark. When $\sup A$ exists, it is unique. Similarly, when $\inf A$ exists, it is unique.

3.2. True or False

Questions. Determine, giving reasons, whether the following statements are true or false.

(1) If a relation is not symmetric, then it is anti-symmetric.
(2) A relation is either symmetric or anti-symmetric.
(3) Let $\mathcal{R}$ be a relation on some non-empty set X. If $\mathcal{R}$ is both symmetric and anti-symmetric, then $x \neq y$ implies that $x\not\mathcal{R}y$ (where $x, y \in X$).
(4) Let X be a non-empty set. If $\mathcal{R}$ is an anti-symmetric relation on X, and for any $x, y \in X$, $x\mathcal{R}y$ and $x \neq y$, then $y\not\mathcal{R}x$.

(5) Let X be a set and $\mathcal{R}$ be a relation supposed *symmetric* and *transitive*. Let $x, y \in X$ be such that $x\mathcal{R}y$. By symmetry, we have $y\mathcal{R}x$ and by transitivity we get $x\mathcal{R}x$, i.e. $\mathcal{R}$ is reflexive. Hence, a symmetric and transitive relation is necessarily reflexive.

Answers.

(1) False! The negation of the definition a symmetric relation just differs from the one of an anti-symmetric relation.

(2) False. A relation may be symmetric and anti-symmetric simultaneously. For instance, one may just consider a relation $\mathcal{R}$ defined on say $\mathbb{R}$, by $x\mathcal{R}y \Leftrightarrow x = y$. Alternatively, you may consider any relation on a singleton.

(3) True. Indeed, if $x\mathcal{R}y$, then given the fact that the relation is symmetric, we get $y\mathcal{R}x$ too. Since $\mathcal{R}$ is also anti-symmetric, we obtain $x = y$.

(4) True. This is just a reformulation of the definition of an anti-symmetric relation using the contrapositive.

(5) False! For example, define a relation $\mathcal{R}$ on $\mathbb{R}$ by: $x\mathcal{R}y \Leftrightarrow xy \neq 0$. Then $\mathcal{R}$ is both symmetric and transitive, yet it is not reflexive since for $x = 0$, it is seen that $x\not\mathcal{R}x$.

3.3. Exercises with Solutions

Exercise 3.3.1. Is the relation $\mathcal{R}$ defined on $\mathbb{Z}$ by

$$x\mathcal{R}y \Longleftrightarrow x = -y$$

reflexive? Is it symmetric? Is it anti-symmetric? Is it transitive?

Solution 3.3.1.

(1) $\mathcal{R}$ is not reflexive: If it were, we would have

$$\forall x \in \mathbb{Z} : x\mathcal{R}x,$$

i.e.

$$\forall x \in \mathbb{Z} : x = -x.$$

But

$$\exists x = 1 \in \mathbb{Z} \text{ such that } 1 = x \neq -x = -1.$$

Hence, $\mathcal{R}$ is not reflexive.

(2) $\mathcal{R}$ is symmetric because for all $x, y \in \mathbb{Z}$:

$$x\mathcal{R}y \Longleftrightarrow x = -y \Longrightarrow y = -x \Longleftrightarrow y\mathcal{R}x.$$

(3) $\mathcal{R}$ is not anti-symmetric because:

$$\exists 1, -1 \in \mathbb{Z} : 1\mathcal{R}(-1) \text{ and } (-1)\mathcal{R}1 \text{ yet } 1 \neq -1.$$

(4) $\mathcal{R}$ is not transitive: For example,

$$\exists 1, -1 \in \mathbb{Z} : \; 1\mathcal{R}(-1) \text{ and } (-1)\mathcal{R}1 \text{ but } 1\cancel{\mathcal{R}}1.$$

Exercise 3.3.2. On the set $A = \{1, 2, 3, 4\}$, we define a relation $\mathcal{R}$ as follows:

$$1\mathcal{R}1, \; 2\mathcal{R}2, \; 2\mathcal{R}3, \; 3\mathcal{R}2, \; 4\mathcal{R}2 \text{ and } 4\mathcal{R}4.$$

Is the relation $\mathcal{R}$ reflexive? Symmetric? Anti-symmetric? Transitive?

Solution 3.3.2.

(1) $\mathcal{R}$ is not reflexive for $3\cancel{\mathcal{R}}3$.
(2) $\mathcal{R}$ is not symmetric since $4\mathcal{R}2$ and $2\cancel{\mathcal{R}}4$!
(3) $\mathcal{R}$ is not anti-symmetric as

$$2\mathcal{R}3 \text{ and } 3\mathcal{R}2 \text{ yet } 2 \neq 3.$$

(4) $\mathcal{R}$ is not transitive because

$$4\mathcal{R}2 \text{ and } 2\mathcal{R}3 \text{ yet } 4\cancel{\mathcal{R}}3.$$

Exercise 3.3.3. Decide whether each of the following relations are reflexive, symmetric, anti-symmetric or transitive:

(1) On $\mathbb{R}$, define:

$$x\mathcal{R}y \iff e^x \leq e^y.$$

(2) The relation in this case is defined as:

$$\Delta\mathcal{R}\Delta' \iff \Delta \perp \Delta'$$

for all lines Δ and Δ' in the plane.
(3) On $\mathbb{R}$, define:

$$x\mathcal{R}y \iff \cos^2 y + \sin^2 x = 1.$$

Solution 3.3.3.

(1) (a) $\mathcal{R}$ is reflexive because

$$\forall x \in \mathbb{R} : \; e^x \leq e^x.$$

(b) $\mathcal{R}$ is not symmetric because

$$\exists 0, 1 \in \mathbb{R} : \; e^0 = 1 \leq e^1 \text{ yet } e^1 \not\leq e^0.$$

(c) $\mathcal{R}$ is anti-symmetric because:

$$\forall x, y \in \mathbb{R} : \begin{cases} x\mathcal{R}y \\ y\mathcal{R}x \end{cases} \iff \begin{cases} e^x \leq e^y \\ e^y \leq e^x \end{cases} \implies e^x = e^y \iff x = y.$$

(d) $\mathcal{R}$ is transitive because

$$\forall x, y, z \in \mathbb{R} : \begin{cases} x\mathcal{R}y \\ y\mathcal{R}z \end{cases} \iff \begin{cases} e^x \leq e^y \\ e^y \leq e^z \end{cases} \implies e^x \leq e^z \iff x\mathcal{R}z.$$

(2) (a) $\mathcal{R}$ is not reflexive because a line cannot be orthogonal (we also say perpendicular) to itself!

(b) $\mathcal{R}$ is symmetric because if a line Δ is orthogonal to another line Δ', then it is patent that Δ' too is orthogonal to Δ.

(c) $\mathcal{R}$ is not anti-symmetric. Indeed, we can have two distinct orthogonal lines Δ and Δ'. For instance, the equations

$$y = -x + 1 \text{ and } y = x + 2$$

define two distinct lines which are orthogonal.

(d) The relation is not transitive. Let's give a counterexample. We need three lines Δ, Δ' and Δ'' such that

$$\Delta \perp \Delta' \text{ and } \Delta' \perp \Delta'' \text{ but } \Delta \not\perp \Delta''.$$

Consider

$$(\Delta) : y = -x + 1, \ (\Delta') : y = x + 2 \text{ and } (\Delta'') : y = -x - 1.$$

Then

$$\Delta \perp \Delta' \text{ and } \Delta' \perp \Delta'' \text{ but } \Delta \not\perp \Delta''$$

(Δ and Δ'' are even parallel!). Hence, $\perp$ is not transitive.

(3) (a) The relation $\mathcal{R}$ is reflexive. Let $x \in \mathbb{R}$. The following trigonometric formula is well-known

$$\cos^2 x + \sin^2 x = 1.$$

It is equivalent to $x \mathcal{R} x$, i.e. $\mathcal{R}$ is reflexive.

(b) $\mathcal{R}$ is symmetric: To see this, let $x, y \in \mathbb{R}$ be such that $x \mathcal{R} y$, i.e. $\cos^2 y + \sin^2 x = 1$. Since

$$\forall x, y \in \mathbb{R} : \ \cos^2 y + \sin^2 y = 1 \text{ and } \cos^2 x + \sin^2 x = 1,$$

it follows that

$$\cos^2 y + \sin^2 x = 1 \iff \cos^2 y + \sin^2 x = \cos^2 y + \sin^2 y$$

or $\sin^2 y = \sin^2 x$. Likewise

$$\cos^2 y + \sin^2 x = 1 \iff \cos^2 y + \sin^2 x = \cos^2 x + \sin^2 x$$

yields $\cos^2 y = \cos^2 x$. Finally, we get

$$\cos^2 y + \sin^2 x = 1 \iff \cos^2 x + \sin^2 y = 1 \iff y \mathcal{R} x,$$

i.e. $\mathcal{R}$ is symmetric.

(c) $\mathcal{R}$ is not anti-symmetric: Recall that this signifies that

$$\exists x, y \in \mathbb{R} : x\mathcal{R}y \text{ and } y\mathcal{R}x \text{ but } x \neq y.$$

For instance, $0\mathcal{R}(2\pi)$ because

$$\cos^2(2\pi) + \sin^2 0 = 1 + 0 = 1,$$

and $(2\pi)\mathcal{R}0$ for

$$\cos^2 0 + \sin^2(2\pi) = 1 + 0 = 1,$$

and yet $0 \neq 2\pi$.

(d) $\mathcal{R}$ is transitive: Let $x, y, z \in \mathbb{R}$ and suppose that $x\mathcal{R}y$ and $y\mathcal{R}z$, i.e.

$$\cos^2 y + \sin^2 x = 1 \text{ and } \cos^2 z + \sin^2 y = 1.$$

Hence

$$\cos^2 y + \sin^2 x + \cos^2 z + \sin^2 y = 1 + 1 = 2.$$

But

$$\sin^2 x + \cos^2 z + \underbrace{\cos^2 y + \sin^2 y}_{=1} = 2,$$

and so

$$\cos^2 z + \sin^2 x = 1,$$

i.e. $x\mathcal{R}z$, i.e. $\mathcal{R}$ is transitive.

Exercise 3.3.4. Let E be the set of prime numbers greater than or equal to 3. Let $\mathcal{R}$ be a relation defined over E by:

$$n\mathcal{R}m \iff \frac{n+m}{2} \in E.$$

(1) Do we have $3\mathcal{R}5$? $3\mathcal{R}11$?
(2) Is $\mathcal{R}$ reflexive? Symmetric? Anti-symmetric? Transitive?

Solution 3.3.4.

(1) We do not have $3\mathcal{R}5$ because $3\mathcal{R}5 \iff (3+5)/2 \in E$ and

$$\frac{3+5}{2} = \frac{8}{2} = 4 \notin E.$$

We do have $3\mathcal{R}11$ because

$$\frac{3+11}{2} = 7 \in E.$$

(2) (a) $\mathcal{R}$ is reflexive because

$$\forall n \in E : n\mathcal{R}n, \text{ since } \frac{n+n}{2} = n \in E.$$

(b) $\mathcal{R}$ is symmetric as

$$\forall n, m \in E : \ n\mathcal{R}m \Longrightarrow m\mathcal{R}n$$

since $(n + m)/2 = (m + n)/2 \in E$.

(c) $\mathcal{R}$ is not anti-symmetric: First, 3 and 7 are in E. Besides, $3\mathcal{R}7$ since $(3 + 7)/2 = 5 \in E$, and $7\mathcal{R}3$ as $(7 + 3)/2 = 5$, and yet $3 \neq 7$.

(d) $\mathcal{R}$ is not transitive because for $3,7,11 \in E$, we have $11\mathcal{R}3$ and $3\mathcal{R}7$, but still $11\not\mathcal{R}7$ since $(11 + 7)/2 = 9 \notin E$.

Exercise 3.3.5. Let $f : \mathbb{R} \to \mathbb{R}$ be a function. We define on $\mathbb{R}$ a relation as follows:

$$x\mathcal{R}y \Longleftrightarrow f(x) = f(y).$$

(1) Show that $\mathcal{R}$ is an equivalence relation.

(2) Give a condition on f so that $\mathcal{R}$ becomes an order relation.

Solution 3.3.5.

(1) That $\mathcal{R}$ is an equivalence relation is left to interested readers.

(2) In order that $\mathcal{R}$ be an order relation, it has to be anti-symmetric, given that it is already reflexive and transitive (by the previous question). Recall that $\mathcal{R}$ is anti-symmetric if for all $x, y \in \mathbb{R}$, $x\mathcal{R}y$ and $y\mathcal{R}x$, we obtain $x = y$. Put differently, for all $x, y \in \mathbb{R}$, we must have

$$f(x) = f(y) \Longrightarrow x = y.$$

It then becomes clear that the injectivity of f resolves this issue.

Exercise 3.3.6. On $\mathbb{Z}$, we define a relation $\mathcal{R}$ by:

$$x\mathcal{R}y \Longleftrightarrow x - y \text{ is divisible by } 4.$$

Show that $\mathcal{R}$ is an equivalence relation, then find all equivalence classes.

Remark. Recall that $x - y$ is divisible by p $(p \in \mathbb{N})$ may be stated as: x is congruent to y modulo p. In symbols, $x \equiv y(\text{mod } p)$.

Solution 3.3.6. We show that $\mathcal{R}$ is reflexive. Let $x \in \mathbb{Z}$. Then $x\mathcal{R}x$ for the simple reason that $x - x = 0$ is divisible by 4.

To show $\mathcal{R}$ is symmetric, let $x, y \in \mathbb{Z}$ and assume that $x\mathcal{R}y$, that is, $x - y$ is divisible by 4. This means that $x - y = 4k$ for some $k \in \mathbb{Z}$. Hence $y - x = -4k = 4(-k)$. Since $(-k) \in \mathbb{Z}$, we see that $y - x$ is divisible by 4, whereby $y\mathcal{R}x$.

To show that $\mathcal{R}$ is transitive, let $x, y, z \in \mathbb{Z}$ and assume that $x\mathcal{R}y$ and $y\mathcal{R}z$. This signifies that $x - y = 4k$ and $y - z = 4k'$ for certain

$k, k' \in \mathbb{Z}$. Hence

$$x - z = x - y + y - z = 4(k + k').$$

Since $k + k' \in \mathbb{Z}$, it follows that $x - z$ is divisible by 4 or that $\mathcal{R}$ is transitive. Thus, $\mathcal{R}$ is an equivalence relation.

To find all equivalence classes, we first find equivalence classes of some particular elements. Let's start with 0 (for example). We have

$$\dot{0} = \{x \in \mathbb{Z} : x\mathcal{R}0\} = \{x \in \mathbb{Z} : x \text{ is divisible by } 4\} = \dot{0} = \{4k : k \in \mathbb{Z}\}.$$

We also have

$$\dot{1} = \{x \in \mathbb{Z} : x\mathcal{R}1\} = \{x \in \mathbb{Z} : x{-}1 \text{ is divisible by } 4\} = \{4k{+}1 : k \in \mathbb{Z}\},$$

and

$$\dot{2} = \{4k + 2 : k \in \mathbb{Z}\}, \quad \dot{3} = \{4k + 3 : k \in \mathbb{Z}\}.$$

Now, if we calculate the equivalence class of 4, we get back the equivalence class of 0, i.e. $\dot{4} = \dot{0}$. Also, $\dot{5} = \dot{1}$, etc. (i.e. no new classes are obtained).

It is easy to see that each integer must belong to one and only one of the equivalence classes: $\dot{0}$, $\dot{1}$, $\dot{2}$ and $\dot{3}$ (remember that constitute a partition of $\mathbb{Z}$).

Remark. The equivalence classes in this case, and as is customary, are written as

$$\mathbb{Z}_4 = \mathbb{Z}/4\mathbb{Z} = \{\dot{0}, \dot{1}, \dot{2}, \dot{3}\}.$$

More generally, if we defined a relation $\mathcal{R}$ by $x\mathcal{R}y$ iff $x{-}y$ is divisible by n, $n \geq 2$, then $\mathcal{R}$ may be shown to be an equivalence relation whose equivalence classes are given by:

$$\mathbb{Z}_n = \mathbb{Z}/n\mathbb{Z} = \{\dot{0}, \dot{1}, \cdots, \dot{(n-1)}\}.$$

Exercise 3.3.7. We define on $\mathbb{R}$ a relation $\mathcal{R}$ as follows:

$$x\mathcal{R}y \Longleftrightarrow x^2 - y^2 = x - y.$$

(1) Show that $\mathcal{R}$ is an equivalence relation.
(2) Let $x \in \mathbb{R}$. Find its equivalence class. How many elements are there in this class?

Solution 3.3.7.

(1) We can check each of the properties of an equivalence relation as done in many cases above, or alternatively we can directly use the result of Exercise 3.3.5 by observing that

$$x^2 - y^2 = x - y \Longleftrightarrow x^2 - x = y^2 - y,$$

then we take $f(x) = x^2 - x$.

(2) Let $x \in \mathbb{R}$. Remember that

$$\dot{x} = \{y \in \mathbb{R} : x\mathcal{R}y\}.$$

To find $\dot{x}$, we need to find all y which satisfy

$$x^2 - y^2 = x - y.$$

We have

$$x^2 - y^2 = x - y \Longleftrightarrow (x-y)(x+y) = x - y \Longleftrightarrow (x-y)(x+y-1) = 0,$$

and so

$$y = x \text{ or } y = 1 - x.$$

Therefore,

$$\dot{x} = \{x, 1 - x\}.$$

In the end, card $\dot{x}$ is constituted of two elements *unless* $x = 1 - x$, i.e. $x = \frac{1}{2}$. To recap,

$$\dot{x} = \begin{cases} \{x, 1-x\}, & \text{if } x \neq \frac{1}{2}, \\ \{\frac{1}{2}\}, & \text{otherwise.} \end{cases}$$

Exercise 3.3.8. On $\mathbb{Z}$, we define an equivalence relation $\mathcal{R}$ by:

$$x\mathcal{R}y \Longleftrightarrow x + y \text{ is even.}$$

(1) Show that $\mathcal{R}$ is an equivalence relation.
(2) What are the equivalence classes of this relation?

Solution 3.3.8.

(1) (a) Let $x \in \mathbb{Z}$. Since $x + x = 2x$ is always even, $\mathcal{R}$ is reflexive.
   (b) Let $x, y \in \mathbb{Z}$. The obvious fact that $x + y$ is even if and only if $y + x$ is so, amply means that $\mathcal{R}$ is symmetric.
   (c) The relation $\mathcal{R}$ is transitive. To prove this, let $x, y, z \in \mathbb{Z}$ and assume that $x\mathcal{R}y$ and $y\mathcal{R}z$, i.e. $x + y$ and $y + z$ are even. So, there exist $n, m \in \mathbb{Z}$ such that $x + y = 2n$ and $y + z = 2m$. Thus,

$$x + y + y + z = 2n + 2m \Longrightarrow x + z = 2\underbrace{(n + m - y)}_{\in \mathbb{Z}},$$

   i.e. $x + z$ is even, that is, $x\mathcal{R}z$.

(2) Observe that with respect to $\mathcal{R}$ here, any even number is in relation with another even number, as is any odd number with another odd number. More precisely, we have:

$$\dot{0} = \{y \in \mathbb{Z} : 0\mathcal{R}z\} = \{y \in \mathbb{Z} : 0 + y \text{ even }\} = \{0, \pm 2, \pm 4, \cdots\}$$

and

$$\dot{1} = \{y \in \mathbb{Z} : 1\mathcal{R}z\} = \{y \in \mathbb{Z} : 1 + y \text{ even }\} = \{\pm 1, \pm 3, \pm 5, \cdots\}.$$

These two classes patently cover all the integers without any overlap. Since the union of these last two sets gives $\mathbb{Z}$ and their intersection is empty, these two classes constitute a partition of $\mathbb{Z}$. Since the equivalence classes always form a partition of the total set, $\dot{0}$ and $\dot{1}$ are the *only* equivalence classes with respect to this equivalence relation.

Exercise 3.3.9. Over $\mathbb{R}^2$, define a relation, noted $\precsim$, by:

$$(x, y) \precsim (x', y') \Longleftrightarrow x \leq x' \text{ and } y \leq y'.$$

(1) Show that $\precsim$ is an order relation.
(2) Is this a total order?
(3) Let $A = \{(x, y) \in \mathbb{R}^2 : x^2 + y^2 \leq 1\}$. Find the upper bounds of A with respect to $\precsim$. What is $\sup A$? Does $\max A$ exist?

Solution 3.3.9.

(1) (a) The relation $\precsim$ is reflexive. Indeed,

$$\forall (x, y) \in \mathbb{R}^2 : (x, y) \precsim (x, y)$$

because

$$\forall (x, y) \in \mathbb{R}^2 : x \leq x \text{ and } y \leq y.$$

   (b) The relation $\precsim$ is anti-symmetric: Let $(x, y), (x', y') \in \mathbb{R}^2$ and suppose $(x, y) \precsim (x', y')$ and $(x', y') \precsim (x, y)$. Then,

$$x \leq x', y \leq y', x' \leq x, \text{ and } y' \leq y.$$

   So, $x = x'$, $y = y'$, that is, $(x, y) = (x', y')$. Hence, $\precsim$ is anti-symmetric.

   (c) The relation $\precsim$ is transitive: Let $(x, y), (x', y'), (x'', y'') \in \mathbb{R}^2$ and assume that $(x, y) \precsim (x', y')$ and $(x', y') \precsim (x'', y'')$, i.e. $x \leq x'$, $y \leq y'$, $x' \leq x''$ and $y' \leq y''$. Thus,

$$x \leq x'' \text{ and } y \leq y'', \text{ i.e. } (x, y) \precsim (x'', y'').$$

(2) The order is not total. If it were, for instance $(1, 2)$ and $(3, 0)$ would be comparable with respect to $\precsim$. This is not the case because

$$(1, 2) \not\precsim (3, 0) \ and \ (3, 0) \not\precsim (1, 2)$$

since

$$(1 \leq 3 \text{ but } 2 \not\leq 0) \ and \ (3 \not\leq 1 \text{ but } 0 \leq 2).$$

(3) Recall that with respect to $\precsim$, an upper bound $(m, M) \in \mathbb{R}^2$ verifies

$$\forall (x, y) \in A : (x, y) \precsim (m, M),$$

i.e.
$$\forall (x, y) \in A : \ x \leq m \text{ and } y \leq M.$$
Let's find the set of upper bounds. Suppose that (m, M) is an upper bound of A, then
$$\forall (x, y) \in A : \ x \leq m \text{ and } y \leq M.$$
In particular, for $(0, 1)$ (which belongs to A) we have:
$$(0, 1) \precsim (m, M), \text{ i.e. } (0 \leq m \text{ and } 1 \leq M).$$
Similarly, since $(1, 0) \in A$, we get:
$$(1, 0) \precsim (m, M) \text{ and so } (1 \leq m \text{ and } 0 \leq M).$$
From these two observations, we see that
$$m \geq 1 \text{ and } M \geq 1.$$
Conversely, let's show that if $m \geq 1$ and $M \geq 1$, then (m, M) is an upper bound of A. Let $(x, y) \in A$, i.e. $x^2 + y^2 \leq 1$. So, $x \leq 1 \leq m$ and $y \leq 1 \leq M$. The set of upper bounds is then given by:
$$\{(m, M) \in \mathbb{R}^2 : \ m \geq 1, M \geq 1\}.$$
The least upper bound of A, that is, $\sup A$ is $(1, 1)$. In the end, $\max A$ does not exists because $\sup A \notin A$.

Exercise 3.3.10. Find the upper and lower bounds of B with respect to $\precsim$ (the order relation defined in Exercise 3.3.9)
$$B = \{(2, 3), (2, 1)\}.$$
What is $\sup B$? What is $\inf B$? Does $\max B$ exist? Does $\min B$ exist?

Solution 3.3.10. Let (m, M) be an upper bound of B, so
$$\forall (x, y) \in B : \ x \leq m \text{ and } y \leq M.$$
That is
$$2 \leq m, \ 3 \leq M \text{ and } 2 \leq m, \ 1 \leq M.$$
So, $2 \leq m$ and $3 \leq M$. Hence, the set of upper bounds is given by:
$$\{(m, M) \in \mathbb{R}^2 : \ m \geq 2, M \geq 3\}.$$
The supremum of B, i.e. $\sup B$ is equal to $(2, 3)$, and we clearly see that $\sup B \in B$. Thus, $\max B = (2, 3)$.
Let (m', M') be a lower bound of B, i.e.
$$\forall (x, y) \in B : \ m' \leq x \text{ and } M' \leq y.$$
So, the set of lower bounds is:
$$\{(m', M') \in \mathbb{R}^2 : \ m' \leq 2, M' \leq 1\}.$$

The infimum of B, i.e. inf B is equal to $(2,1)$. Since $(2,1)$ belongs to B, we conclude that min $B = (2,1)$.

Exercise 3.3.11. Let $\mathcal{R}$ be the relation defined on $\mathbb{N}$ by:

$$x\mathcal{R}y \iff x \text{ divides } y.$$

(1) Prove that $\mathcal{R}$ is an order relation.
(2) Is a total order?
(3) Let

$$A = \{2,3,5\}.$$

Find the sets of upper and lower bounds of A. Determine sup A and inf A. What about max A and min A?
(4) The same questions for the set B given by

$$B = \{4,8,16\}.$$

Solution 3.3.11.

(1) Recall that x divides y in $\mathbb{N}$ if and only if there exists a certain $n \in \mathbb{N}$ such that $y = nx$.
   (a) The relation is reflexive because we can divide any natural number by itself.
   (b) To show that $\mathcal{R}$ is anti-symmetric, let $x, y \in \mathbb{N}$ and suppose $x\mathcal{R}y$ and $y\mathcal{R}x$, i.e. x divides y and y divides x, i.e. there exist $n, m \in \mathbb{N}$ such that $y = nx$ and $x = my$. Hence,

   $$y = nmy \implies 1 = nm \text{ because } y \neq 0.$$

   However, the *only* natural integers n and m satisfying $nm = 1$ are $n = 1$ and $m = 1$. Hence $x = y$, i.e. the relation is indeed anti-symmetric.
   (c) Let $x, y, z \in \mathbb{N}$, and assume $x\mathcal{R}y$ and $y\mathcal{R}z$, i.e. there are $n, m \in \mathbb{N}$ such that $y = nx$ and $z = my$. So $z = mnx$, i.e. x divides z, that is $x\mathcal{R}z$, i.e. $\mathcal{R}$ is transitive, and so we have shown that the relation $\mathcal{R}$ is an order relation.
(2) The order is not total. We have to find two positive integers x and y which are not comparable, i.e. such that *neither* x divides y, *nor* y divides x. For example, 3 and 5 are not comparable as 3 does not divide 5 and 5 does not divide 3.
(3) Let's find the sets of upper and lower bounds of A. If M is an upper bound of A, then

$$\forall x \in A : x \text{ divides } M.$$

It is clear that M must be a multiple of 2, 3 and 5 at the same time. Hence, the set of upper bounds of A consists of all multiples of 2, 3 and 5 simultaneously, i.e. multiples of 30.

Now, if m is a lower bound of A, then

$$\forall x \in A : m \text{ divides } x.$$

So m must divide 2, 3 and 5 at the same time, and so it is clear that the set of lower bounds is reduced to $\{1\}$.

Thus, the supremum of A is:

$$\sup A = \text{lcm}\{2, 3, 5\} = 30.$$

Since $30 \notin A$, it is seen that $\max A$ does not exist!

Finally, clearly $\inf A = 1$, and since $1 \notin A$, we infer that $\min A$ does not exist.

(4) Applying the method of the previous question we find that the set of upper bounds consists of multiples of 4, 8 and 16 at once (i.e. the multiples of 16). So

$$\sup B = \text{lcm}\{4, 8, 16\} = 16.$$

Since $16 \in B$, we conclude that $\max B$ exists and it is equal to 16.

The set of lower bounds of B is the set of divisors of 4, 8 and 16 simultaneously, i.e. it is formed of numbers 1, 2 and 4. The largest of them is the greatest common divisor, noted gcd, i.e.

$$\inf B = \gcd\{4, 8, 16\} = 4.$$

It is also a minimum because $4 \in B$, i.e. $\min B = 4$.

Exercise 3.3.12. Let X be a non-empty set and let $E = \mathcal{P}(X)$ be the power set of X. We define a binary relation $\mathcal{R}$ over E as follows:

$$A\mathcal{R}B \iff A \subset B.$$

(1) Show that $\mathcal{R}$ is an order relation over E.
(2) Is the order total?
(3) Set $X = \{3, 4\}$. Find the upper bounds, the lower bounds, the supremum, the infimum, the biggest element and the smallest element (if they exist) of $F = \{\{3\}\}$.
(4) The same questions with $X = \mathbb{R}$ and $F = \{\mathbb{R}_-^*, \mathbb{R}_+^*\}$ where $\mathbb{R}_-^* = (-\infty, 0)$.

Solution 3.3.12.

(1) (a) Let $A \in E$, then $A \subset A$. Hence, $\mathcal{R}$ is reflexive.

(b) Let $A, B \in E$ be such that $A\mathcal{R}B$ and $B\mathcal{R}A$, i.e. $A \subset B$ and $B \subset A$. So, we get $A = B$. So $\mathcal{R}$ is anti-symmetric.

(c) Finally, $\mathcal{R}$ is transitive because

$$\forall A, B, C \in E : (A \subset B \text{ and } B \subset C \Longrightarrow A \subset C).$$

Therefore, $\mathcal{R}$ is an order relation.

(2) The order is not total. To see why, we need to find $A, B \in E$ such that

$$A \not\subset B \text{ and } B \not\subset A.$$

Since we do not have much information about X, the best counterexample perhaps is to take any $A \subset X$ (with $A \neq \varnothing$ and $A \neq X$) and $B = A^c$, i.e. the complement of A. So it is obvious that

$$A \not\subset A^c \text{ and } A^c \not\subset A.$$

(3) We have

$$E = \mathcal{P}(X) = \{\varnothing, \{3\}, \{4\}, \{3, 4\}\}.$$

Remember that $M \in E$ is an upper bound of $F = \{\{3\}\}$ if

$$\{3\} \subset M,$$

whereby the set of upper bounds is given by

$$\{\{3\}, \{3, 4\}\}.$$

Hence $\sup F = \{3\}$, and since $\{3\} \in F = \{\{3\}\}$, we deduce that $\max F$ exists and it is equal to $\{3\}$.

On the other hand, the set of lower bounds is given by:

$$\{\varnothing, \{3\}\},$$

and so $\inf F = \varnothing$, and $\min F$ does not exist because $\varnothing \notin F$.

(4) A set M is an upper bound of F if

$$\mathbb{R}_-^* \subset M \text{ and } \mathbb{R}_+^* \subset M.$$

It is then seen that the set of upper bounds is reduced to $\{\mathbb{R}^*, \mathbb{R}\}$. The set of lower bounds is reduced to $\{\varnothing\}$. Whence,

$$\sup F = \mathbb{R}^*, \ \inf F = \varnothing.$$

Since $\mathbb{R}^*, \varnothing \notin F$, we conclude that neither $\max F$ nor $\min F$ exists.

3.4. Supplementary Exercises

Exercise 3.4.1. Define a relation $\mathcal{R}$ over $E = \{0, 1, 2\}$ as follows

$$0\mathcal{R}0, \ 0\mathcal{R}1, \ 1\mathcal{R}1 \text{ and } 2\mathcal{R}2.$$

Is $\mathcal{R}$ an order relation on E?

The same question with

$$0\mathcal{R}0, \ 0\mathcal{R}1, \ 1\mathcal{R}1, 1\mathcal{R}2 \text{ and } 2\mathcal{R}2.$$

Exercise 3.4.2. A binary relation $\mathcal{R}$ over a set E is said to be circular when

$$\forall x, y, z \in E : \ (x\mathcal{R}y \text{ and } y\mathcal{R}z) \implies z\mathcal{R}x.$$

Show that a circular relation that is reflexive is always an equivalence relation.

Exercise 3.4.3. Is the relation $\mathcal{R}$ defined on $\mathbb{R}$ by

$$x\mathcal{R}y \iff |x| \leq |y|$$

a total order?

Exercise 3.4.4. Determine whether the following relations (defined on $\mathbb{R}$) are order or equivalence relations:

(1) $x\mathcal{R}y \iff [x] \leq [y]$,
(2) $x\mathcal{R}y \iff [x] = [y]$,
(3) $x\mathcal{R}y \iff \cos x = \cos y$,

where $[\cdot]$ is the usual greatest integer function.

Exercise 3.4.5. Let $\mathcal{R}$ be a relation defined over, $\mathbb{N} \times \mathbb{N}$ by

$$(a, b)\mathcal{R}(c, d) \iff ad = bc.$$

Prove that $\mathcal{R}$ is an equivalence relation on $\mathbb{N} \times \mathbb{N}$.

Exercise 3.4.6. Consider the relation defined in Exercise 3.3.10, then set

$$A = \{(1, 2), (0, 4), (3, 1)\}.$$

Find the upper and lower bounds of A. Find also $\sup A$, $\inf A$ and $\max A$ and $\min A$ (if they exist).

Exercise 3.4.7. Define on $\mathbb{N}$ a relation $\mathcal{R}$ by

$$n\mathcal{R}m \iff \exists p \in \mathbb{N} : \ m = n^p.$$

(1) Check that $\mathcal{R}$ is an order relation on $\mathbb{N}$.
(2) Let $A = \{2, 4, 16\}$. Find the upper and lower bounds of A. Find $\sup A$, $\inf A$ and $\max A$ and $\min A$ (if they exist).
(3) The same questions with $B = \{4, 8\}$.

CHAPTER 4

Groups

4.1. Basics

4.1.1. Groups.

DEFINITION 4.1.1. Say that a non-empty set G with a operation $*$ is a group if $*$ satisfies the following laws:

(1) For all $x, y \in G$: $x * y \in G$ (closure law). Alternatively, we may say that "$*$" is a binary operation on G, i.e. it defines a map $* : G \times G \to G$ by $(x, y) \mapsto x * y$.

(2) For all $x, y, z \in G$: $(x * y) * z = x * (y * z)$ (associative law).

(3) There is an $e \in G$ such that for all $x \in G$: $e * x = x * e = x$ (identity law).

(4) For all $x \in G$, there is an $x' \in G$ such that: $x * x' = x' * x = e$ (inverse law). The element x' is called the inverse of x.

A group with a binary operation $*$ is designated by $(G, *)$.

If $(G, *)$ is a group and $x * y = y * x$ for all $x, y \in G$ (commutative law), then $(G, *)$ is called a commutative or abelian group.

Remark. The inverse of an element x in some group G may be denoted by x^{-1} or $-x$ depending on the notation of the binary operation.

Remark. There are different notations for binary operations. For example: $*$, $\circ$, $\bullet$, $\perp$, $\oplus$, $\otimes$, $\cdot$, $+$, $\times$, etc. Readers should also bear in mind that the operations $+$, $\times$ or $\cdot$ do not necessarily stand for the usual addition and multiplication of usual numbers say.

Remark. When there is no risk of confusion, we may write xy or $x \cdot y$ instead of $x * y$, even when we are not dealing with the usual multiplication of numbers.

We will see many examples of non-commutative groups. However, and for some purposes, we do not need the whole operation $*$ to be commutative, it suffices to have that property true for two given elements. This notion is important enough to be singled out:

DEFINITION 4.1.2. Let $(G, *)$ be a group. We say that $a, b \in G$ commute provided that

$$a * b = b * a.$$

Now, we pass to some basic examples. We start with examples about the closure axiom.

EXAMPLES 4.1.1.

(1) The addition and multiplication do not satisfy the closure axiom in the set of irrationals. Counterexamples are already available in the "True or False" Section in Chapter 1.
(2) The subtraction "−" is not a binary operation on $\mathbb{N}$. For instance, $1, 2 \in \mathbb{N}$ but $1 - 2 = -1 \notin \mathbb{N}$
(3) The closure axiom is not satisfied by the usual addition in the set of odd natural numbers. As a counterexample, 3 and 5 are odd numbers whereas their sum, i.e. 8, is not.
(4) The usual addition satisfies the closure axiom in the set of integers.
(5) The usual multiplication satisfies the closure axiom in the set of integers.

Next, we give examples of groups and non-groups.

EXAMPLES 4.1.2.

(1) If "+" denotes the usual addition of numbers, then $(\mathbb{Z}, +)$, $(\mathbb{Q}, +)$, $(\mathbb{R}, +)$ and $(\mathbb{C}, +)$ are abelian groups. The identity element is 0, and the inverse of each x is $-x$. This follows from the axioms of numbers.
(2) If "×" designates the usual multiplication of numbers, then none of $(\mathbb{Z}, \times)$, $(\mathbb{Q}, \times)$, $(\mathbb{R}, \times)$ and $(\mathbb{C}, \times)$ is a group. The identity element is 1, but 0 has no inverse.
(3) $(\mathbb{Q}^*, \times)$, $(\mathbb{R}^*, \times)$ and $(\mathbb{C}^*, \times)$ are all abelian groups. The identity element is 1, and the inverse of each x is $1/x$.
(4) $(\mathbb{Z}^*, \times)$ is still not a group (e.g. what is the inverse of 2?).

Next, we list some basic properties of the identity and inverse elements.

PROPOSITION 4.1.1. *Let $(G, *)$ be a group.*

(1) *The identity element of a group is unique.*
(2) *The inverse of any element of a group is unique as well.*
(3) *The inverse of the inverse of an element is the element itself.*
(4) *If $x, y \in G$, then*

$$(x * y)^{-1} = y^{-1} * x^{-1}.$$

PROPOSITION 4.1.2. *(A proof may consulted in Exercise 4.3.1) Let $(G, *)$ be a group, and let $a, b, x \in G$. Then*

(1) $a * x = b * x \implies a = b$ *(right cancellation law).*

(2) $x * a = x * b \Longrightarrow a = b$ *(left cancellation law)*.

Next we introduce the notion of exponentiation in groups.

DEFINITION 4.1.3. Let $(G, *)$ be a group, with e being its identity element. Let $x \in G$ and let $n \in \mathbb{Z}$. If $n > 0$, define

$$x^n = \underbrace{x * x * \cdots * x}_{n \text{ times}}.$$

If $n = 0$, set $x^0 = e$. When $n < 0$, then setting $n = -m$, where $m \in \mathbb{N}$, we define $x^n = x^{-m} = (x^m)^{-1}$.

Remark. It would be meaningless to define fractional exponents in groups, but students may try to see what goes wrong should they want to give this a go.

The familiar laws of exponentiation do hold for a group.

PROPOSITION 4.1.3. *Let $(G, *)$ be a group, and let $x \in G$. If $n, m \in \mathbb{Z}$, then*

(1) $x^n * x^m = x^{n+m}$.
(2) $(x^n)^m = x^{nm}$.

Remark. We have written $x * x * \cdots * x$ in a similar way as if $*$ was $\times$. The analogous notation with respect to some addition would be nx. Readers should be able to figure out what the other identities in this case are.

4.1.2. Subgroups.
If we ask students to try to divine a definition of subgroup, then they would probably say that it is some subset of the "bigger" group, in which all the laws of a group are satisfied by the same binary operation. Well, this is almost true, as they should add that the subset must not be empty.

However, there are two equivalent definitions more known as subgroups tests. We state the most practical one, namely:

PROPOSITION 4.1.4. *(Subgroup test) Let G be a group with e as its identity element, and let H be a non-empty subset of G. Then H is a subgroup of G if and only if*

$$\forall x, y \in H : xy^{-1} \in H,$$

where as needed and as alluded to before, the expression xy^{-1} must be adapted to the appropriate binary operation (observe in passing that this implies that $e \in H$).

Remark. Readers must not forget to check that H is non-empty. To show H is non-empty, it suffices to find any element which belongs to it. Usually, we check whether $e \in H$. If $e \in H$, then H is non-empty but we cannot decide yet whether H is a subgroup. We still need to see whether the subgroup test holds.

But, if $e \notin H$, then we immediately declare that H is not a subgroup.

Remark. One important virtue of subgroups is that if one needs to show that a given set is a group, then one should try to observe whether this given set may be regarded as a subset of another set, which is known to be a group, and just apply the subgroup test.

4.1.3. Homomorphisms et al.

DEFINITION 4.1.4. Let $(G, *)$ and $(G', \perp)$ be two groups and let $f : G \to G'$ be a function. We say that f is a group homomorphism (or homomorphism of groups) if for all $x, y \in G$, the following holds

$$(x * y) = f(x) \perp f(y).$$

Notice that the words homomorphism or morphism may be used interchangeably. There are, however, other different concepts.

DEFINITION 4.1.5. Let $f : G \to G'$ be a group homomorphism. Then

(1) f is said to be an endomorphism when $G = G'$.
(2) f is called an isomorphism if f is bijective.
(3) Say that f is an automorphism if f is an isomorphism and $G = G'$.

If $f : G \to G'$ is a homomorphism, then we say that G is homomorphic to G'; and if $f : G \to G'$ is an isomorphism, then we say that G is isomorphic to G'.

Here are some standard properties about homomorphisms.

THEOREM 4.1.1. *Let $f : G \to G'$ and $g : G' \to G''$ be two group homomorphisms.*

(1) *If e is the identity element of G and e' is that of G', then $f(e) = e'$.*
(2) *For all $x \in G$: $f(x^{-1}) = (f(x))^{-1}$*
(3) *If H is a subgroup of G, then $f(H)$ is a subgroup of G'.*
(4) *If H' is a subgroup of G', then $f^{-1}(H')$ is a subgroup of G.*
(5) *$g \circ f : G \to G''$ is a group homomorphism.*

DEFINITION 4.1.6. Let $f : G \to G'$ be a group homomorphism.

(1) The kernel of f, noted ker f, is the subset of G defined by:

$$\ker f = \{x \in G : f(x) = e'\}.$$

(2) The range of f, denoted by ran f, is the subset of G' defined by:

$$\operatorname{ran} f = \{f(x) : x \in G\}$$

PROPOSITION 4.1.5. *Let $f : G \to G'$ be a group homomorphism. Then*

(1) ker f *is a subgroup of G.*
(2) ran f *is a subgroup of G'.*
(3) f *is one-to-one if and only if* ker $f = \{e\}$.

Remark. That f is onto if and only if ran $f = G'$ is already known in a more general context.

4.1.4. A word on matrices.

Adding the definition and some basic notions about matrices is primordial. This will very much enrich the present manuscript. We content ourselves to the case of 2×2 matrices, which is just enough for including some interesting examples and problems.

DEFINITION 4.1.7. A real 2×2 (square) matrix A is a table of two rows and two columns defined as

$$A = \begin{pmatrix} a & b \\ c & d \end{pmatrix},$$

where a, b, c, d, called the entries of A, are real. The set of real 2×2 matrices is denoted by $M_2(\mathbb{R})$.

Remark. We may also define matrices of higher order, even matrices whose number of its rows differs from that of its columns. Also, the entries may be allowed to be in $\mathbb{C}$ (or in any other structure).

The next definition furnishes the set $M_2(\mathbb{R})$ a little more.

DEFINITION 4.1.8. Let $A = \begin{pmatrix} a & b \\ c & d \end{pmatrix}$ and $B = \begin{pmatrix} x & y \\ z & t \end{pmatrix}$ be two real matrices. Then

(1) The sum of A and B, denoted by $A+B$ is carried out coordinate-wise, i.e. it is defined as

$$A + B = \begin{pmatrix} a + x & b + y \\ c + z & d + t \end{pmatrix}.$$

(2) The multiplication (or product) of A and B, noted AB, is defined by

$$AB = \begin{pmatrix} ax + bz & ay + bt \\ cx + dz & cy + dt \end{pmatrix}$$

(3) If A is a real matrix and α is in $\mathbb{R}$, then

$$\alpha A := \begin{pmatrix} \alpha a & \alpha b \\ \alpha c & \alpha d \end{pmatrix}.$$

Recall the following result (known to readers already familiar with matrices) which holds for (square) matrices of higher order as well.

PROPOSITION 4.1.6. *Let A, B and C be 2×2 real matrices. The product of matrices is associative, i.e.*

$$(AB)C = A(BC).$$

Also,

$$A(B + C) = AB + AC \text{ and } (A + B)C = AC + BC.$$

The set $M_2(\mathbb{R})$ equipped with the multiplication of matrices is not a group. As readers are probably already aware, a matrix $A = \begin{pmatrix} a & b \\ c & d \end{pmatrix}$ is invertible if and only if its determinant, i.e. $\det A := ad - bc$, is non-zero. So, we give the following definition (which is valid for higher order square matrices as well):

DEFINITION 4.1.9. The general linear group, denoted by $GL_2(\mathbb{R})$, is the "group" of invertible 2×2 real matrices under matrix multiplication. In other words,

$$GL_2(\mathbb{R}) = \left\{ A = \begin{pmatrix} a & b \\ c & d \end{pmatrix} : a, b, c, d \in \mathbb{R}, \det A = ad - bc \neq 0 \right\}.$$

Here is an interesting property about the determinant of a matrix.

PROPOSITION 4.1.7. *Clearly, "det" is a function from $M_2(\mathbb{R})$ into $\mathbb{R}$. If $A, B \in M_2(\mathbb{R})$, then*

$$\det(AB) = \det A \det B.$$

Readers may comfortably show the following result:

PROPOSITION 4.1.8. *The multiplication of two matrices in $GL_2(\mathbb{R})$ is again an element in $GL_2(\mathbb{R})$. The multiplication is associative. If we set $I_2 = \begin{pmatrix} 1 & 0 \\ 0 & 1 \end{pmatrix}$, which is called the identity matrix, we see that for all $A \in GL_2(\mathbb{R})$:*

$$AI_2 = I_2 A = A$$

(hence I_2 is the identity element of $GL_2(\mathbb{R})$ w.r.t. "$\cdot$"). The inverse of $A = \begin{pmatrix} a & b \\ c & d \end{pmatrix} \in GL_2(\mathbb{R})$ is given by

$$A^{-1} = \begin{pmatrix} d/(ad - bc) & -b/(ad - bc) \\ -c/(ad - bc) & a/(ad - bc) \end{pmatrix}.$$

In other words, $GL_2(\mathbb{R})$ is a group under matrix multiplication.

4.2. True or False

Questions. Determine, providing justification, whether the following statements are true or false.

(1) Students have probably already had the opportunity to deal with scalar products on $\mathbb{R}^2$ or $\mathbb{R}^3$. Is the scalar product on $\mathbb{R}^2$ say, a binary operation?

(2) 0 is the identity element with respect to the subtraction ("$-$") over $\mathbb{Z}$.

(3) Subtraction "$-$" over $\mathbb{Z}$ is associative.

(4) Give a set and a binary operation which does not satisfy any of the axioms of an abelian group.

(5) Show that the identity element, when it exists, is unique.

(6) Let X be a non-empty set. Let $G = \mathcal{P}(X)$ be the powerset of X. Consider

$$\forall A, B \in G: \ A * B = A \cup B.$$

We know that the identity element, when it exists, is *unique*. In this case, $\varnothing$ is the identity element. However, each $B \subset A$ may also be perceived as "an identity element" since

$$A \cup B = B \cup A = A, \ \forall A \in G.$$

Explain why this this reasoning is erroneous.

(7) Let $(G, *)$ be a group with an identity element noted e. Let $x \in G$ be such that $x^5 = e$ and $x^2 \neq e$. Then $x^4 = e$.

(8) The multiplication of matrices is commutative.

(9) Let $(G, *)$ be a group containing at least two distinct elements a and b. Let a^{-1} be the inverse of a. Can the inverse of b be equal to a^{-1}?

(10) Give an example of a group $(G, *)$ such that

$$(a * b)^{-1} \neq a^{-1} * b^{-1}$$

for some $a, b \in G$.

(11) If $(G, *)$ is a non-abelian group, then $a*b \neq b*a$ for all $a, b \in G$.

(12) In a non-abelian group $(G, *)$, there are always at least two elements which commute.

(13) A singleton $\{x\}$ may always be given the structure of a group.

(14) Is there a group with two elements only?

(15) A group is a subgroup, and vice versa.

(16) Let G be a group with e as its identity element, and let $H \subset G$ be a subgroup of G, and let H^c be its complement in G. Then H^c can be a subgroup.

(17) All subgroups of an abelian group must themselves be abelian.

(18) Any subgroup of a non-abelian group is itself non-abelian.

(19) Let $(G, *)$ be a non-group, where "$*$" is a binary operation. Then any subset of G is not a group either.

(20) The intersection of two groups is a group.

(21) The union of two groups is a group.

(22) Let f be a group homomorphism. Then $\ker f$ is always reduced to a singleton.

(23) Let $\mathbb{Q}_+^*$ be the set of strictly positive rational numbers. Then $(\mathbb{Q}_+^*, \cdot)$ is isomorphic to $(\mathbb{Q}, +)$.

(24) If $f : (G, *) \to (G', \square)$ is a group isomorphism, then $f^{-1} : (G', \square) \to (G, *)$ is also a group isomorphism.

Answers.

(1) The answer is negative. First, recall that if $u = \begin{pmatrix} x \\ y \end{pmatrix}$ and $u' = \begin{pmatrix} x' \\ y' \end{pmatrix}$ are two vectors in $\mathbb{R}^2$, then their scalar product, noted $u \cdot u'$, is defined by

$$u \cdot u' = xx' + yy'.$$

If "$\cdot$" were a binary operation, it would be defined as a map from $\mathbb{R}^2 \times \mathbb{R}^2$ into $\mathbb{R}^2$. However, $u \cdot u' \in \mathbb{R}$ for all $u, u' \in \mathbb{R}^2$. So, the scalar product is not a binary operation on $\mathbb{R}^2$.

(2) False! If 0 were the identity element of "$-$", then we would have

$$\forall x \in \mathbb{Z} : \quad x - 0 = 0 - x = x.$$

But this is not always true (witness, e.g. $x = 1$).

(3) False! The subtraction is not associative. For instance,

$$(1 - 2) - 3 = -1 - 3 = -4 \neq 1 - (2 - 3) = 1 - (-1) = +2.$$

(4) Consider $\mathbb{Z}$ with the usual subtraction "$-$". Then, "$-$" is not commutative as, e.g. $1 - 2 \neq 2 - 1$. Also, we have just seen that "$-$" is not associative. Moreover, we saw above that 0 is

not an identity element for $(\mathbb{Z}, -)$. In fact, there is no identity element whatsoever. Indeed, if e is such an element, then

$$e - x = x - e = x$$

for every $x \in \mathbb{Z}$. However, the previous equations give $e = 0$ and $e = 2x$! So, there is no identity element, and since the latter does not exist, one cannot even speak of inverses!

(5) Let $(G, *)$ be a group and let e be its identity element. We assume for the sake of contradiction that e' is another identity element in G, and we show that $e = e'$. Since e is an identity element, we have:

$$e' * e = e * e' = e'.$$

Since we have assumed that e' is an identity element as well, we would obtain

$$e' * e = e * e' = e.$$

Consequently, $e = e'$, as wished.

(6) Before clarifying this point, I would like to reassure students that the identity element, when it exists, is always unique.

To see why the given reasoning is untrue, students have to write again the definition of the identity element, i.e.

$$\exists e \in E, \ \forall x \in E : \ e * x = x * e = x.$$

The fact that "$\exists$" is placed before "$\forall$", means that the "e" must be **independent** of x! In the question, B depends of A.

(7) False. Since $x * x \neq e$, it is seen that $x \neq x^{-1}$ and $x \neq e$. On the other hand, $x * x * x * x = e$ gives $x * x * x = x^{-1}$, and so $x * x * x * x \neq e$.

(8) False! For instance, consider

$$A = \begin{pmatrix} 0 & 1 \\ 0 & 0 \end{pmatrix} \text{ and } B = \begin{pmatrix} 1 & 0 \\ 0 & 0 \end{pmatrix}.$$

Then

$$AB = \begin{pmatrix} 0 & 0 \\ 0 & 0 \end{pmatrix} \neq \begin{pmatrix} 0 & 1 \\ 0 & 0 \end{pmatrix} = BA.$$

(9) False. Indeed, if $b^{-1} = a^{-1}$, then $b = a$!

(10) On $GL_2(\mathbb{R})$, let $A = \begin{pmatrix} 0 & 1 \\ 2 & 0 \end{pmatrix}$ and $B = \begin{pmatrix} 0 & 1 \\ 1 & 0 \end{pmatrix}$. Then A and B are both invertible as $\det A = -2$ and $\det B = -1$.

Moreover,

$$AB = \begin{pmatrix} 1 & 0 \\ 0 & 2 \end{pmatrix}, \ A^{-1} = \begin{pmatrix} 0 & 1/2 \\ 1 & 0 \end{pmatrix} \text{ and } B^{-1} = \begin{pmatrix} 0 & 1 \\ 1 & 0 \end{pmatrix}$$

(by a glance at Proposition 4.1.8). Therefore,

$$(AB)^{-1} = \begin{pmatrix} 1 & 0 \\ 0 & 1/2 \end{pmatrix} \neq \begin{pmatrix} 1/2 & 0 \\ 0 & 1 \end{pmatrix} = A^{-1}B^{-1} = \begin{pmatrix} 0 & 1/2 \\ 1 & 0 \end{pmatrix} \begin{pmatrix} 0 & 1 \\ 1 & 0 \end{pmatrix}.$$

(11) Untrue! If $(G, *)$ is non-abelian, then this signifies that

$$\exists a, b \in G : a * b \neq a * b.$$

Other than the latter fact, there are non-abelian groups in which two particular elements commute. For example, the group of 2×2 matrices under the usual multiplication of matrices is not abelian, and yet commuting matrices abundantly exist. For example, the *invertible* matrix $A = \begin{pmatrix} 1 & 0 \\ 0 & -1 \end{pmatrix}$ commutes with the *invertible* matrix $B = \begin{pmatrix} 2 & 0 \\ 0 & -1 \end{pmatrix}$, as may be verified.

(12) True! In order to avoid trivial cases, assume that $\operatorname{card} G \geq 2$. If e is the identity element and if x is another element in G, then e always commues with x because by definition $x*e = e*x$ $(=x)$. Another instance is $x \in G$ and its inverse x^{-1}. Then $x * x^{-1} = x^{-1} * x \ (=e)$.

(13) We all agree that we first require a binary operation, i.e. the closure axiom must be satisfied. Once that is known, we see that the only possible definition of a binary operation in this case is $x * x = x$. Now, $(\{x\}, *)$ is indeed a group (where the identity element is x). Notice in the end that this group is more commonly known as the trivial group.

(14) Yes, and there are plenty of them. Readers may consult, e.g. Exercises 4.3.5 & 4.3.9. See also Exercise 4.3.10 in the case $n = 2$.

(15) True for both ways! A subgroup is patently a group, and a group may be regarded as a subgroup of itself (why not?).

(16) False! In fact, H^c is never a subgroup once H is a subgroup. One reason is that H does contain e, and so $e \notin H^c$, and this prevents H^c from being a subgroup.

(17) True. Let $(G, *)$ be an abelian group, and let H be *any* subgroup of G. Since $*$ is commutative, by definition we know

that
$$\forall x, y \in G : x * y = y * x.$$
Hence, and since $H \subset G$,
$$\forall x, y \in H : x * y = y * x,$$
which shows that $(H, *)$ is abelian too.

(18) False! There are many counterexamples. The most trivial one is to consider any non-abelian group G with e as its identity element. Then $\{e\}$ is an abelian (trivial though) subgroup of G.

As another example, consider $GL_2(\mathbb{R})$ equipped with the multiplication of matrices. Now, consider the set
$$H = \left\{ \begin{pmatrix} a & 0 \\ 0 & b \end{pmatrix} : a, b \in \mathbb{R} \text{ with } ab \neq 0 \right\}.$$

That H is a subgroup of (the non-abelian) $(GL_2(\mathbb{R}), \cdot)$ is left to interested readers to work out details. Now, we show that H is abelian. Let $A = \begin{pmatrix} a & 0 \\ 0 & b \end{pmatrix}$ and $B = \begin{pmatrix} a' & 0 \\ 0 & b' \end{pmatrix}$. Since the usual multiplication on $\mathbb{R}$ is commutative, we may write
$$AB = \begin{pmatrix} aa' & 0 \\ 0 & bb' \end{pmatrix} = \begin{pmatrix} a'a & 0 \\ 0 & b'b \end{pmatrix} = BA,$$
and the proof is complete.

(19) The answer depends on what prevents $(G, *)$ from being a group in the first place. For example, $(\mathbb{R}, \cdot)$ is not a group as, e.g. 0 has no inverse. However, $\{1, -1\}$ equipped with the operation "$\cdot$" is a group as shall be seen in one of the exercises below.

Now, $(\mathbb{N}, +)$ is not a group, and for all $A \subset \mathbb{N}$, $(A, +)$ is never a group.

Finally, if "$*$" is a binary operation and $(G, *)$ is not a group, but it has an identity element, noted e, then $(\{e\}, *)$ is a group!

(20) The intersection of two groups (G and G' say) need not be a group! The main reason is that it is quite conceivable to have $G \cap G' = \varnothing$, i.e. $G \cap G'$ would not even have an identity element. As an explicit counterexample take the intersection of two groups which are not of the same nature (e.g. one group of functions with respect to the composition operation, and the other being some group of numbers endowed with the usual addition). However, the intersection of two subgroups (of a

certain group) is always a subgroup. Readers may consult a
proof in Exercise 4.3.36.

(21) False too. In fact, the union of two subgroups of the same
group may fail to be a group. See Exercise 4.3.36 for a counterexample, as well as a sufficient and necessary condition of
when this is true.

(22) Untrue. Consider $f : \mathbb{R}^* \to \mathbb{R}^*$ defined by $f(x) = x^2$, where
both "$\mathbb{R}^*$s" are equipped with the usual multiplication operation. Then f is a group homomorphism for

$$f(xy) = (xy)^2 = x^2 y^2 = f(x)f(y)$$

for all $x, y \in \mathbb{R}^*$. Now, since 1 is the identity element for
$(\mathbb{R}^*, \cdot)$, the definition of ker f is as follows

$$\ker f = \{x \in \mathbb{R}^* : x^2 = 1\},$$

and so ker $f = \{1, -1\}$ is a doubleton (i.e. a set having two
elements). Observe in the end that since ker $f \neq \{1\}$, f is not
one-to-one.

(23) False! To reach a contradiction, assume that there is an iso-
morphism $f : (\mathbb{Q}, +) \to (\mathbb{Q}_+^*, \cdot)$ with $f(x + y) = f(x) \cdot f(y)$ for
all $x, y \in \mathbb{Q}$. Since f is onto, there is at least an $a \in \mathbb{Q}$ such
that $f(a) = 2$. Hence

$$f(a) = f(a/2 + a/2) = f(a/2) \cdot f(a/2) = [f(a/2)]^2 = 2.$$

Since $f(a/2)$ is a rational number, we know that $[f(a/2)]^2 = 2$
is just impossible over $\mathbb{Q}$. Thus, $(\mathbb{Q}_+^*, \cdot)$ is not isomorphic to
$(\mathbb{Q}, +)$.

(24) True. Let us give a proof. Assume that f is an isomorphism.
Since $f : G \to G'$ is bijective, so is $f^{-1} : G' \to G$. Now,
let $x, y \in G'$. Then, and since f is bijective, $x = f(a)$ and
$y = f(b)$ for some unique $a, b \in G$, and so $a = f^{-1}(x)$ and
$b = f^{-1}(y)$. Since f is a group homomorphism,

$$f(a * b) = f(a) \square f(b).$$

Hence

$$f^{-1}(x \square y) = f^{-1}[f(a) \square f(b)] = f^{-1}[f(a * b)] = a * b = f^{-1}(x) * f^{-1}(y),$$

that is, we have shown that f^{-1} is a group homomorphism.
Thus, f^{-1} is an isomorphism.

4.3. Exercises with Solutions

Exercise 4.3.1. Let $(G, *)$ be a group and let e be its identity element. Let $a, b, x \in G$.

(1) Prove that

$$a * x = b * x \Longrightarrow a = b.$$

(2) Prove that

$$x * a = x * b \Longrightarrow a = b.$$

Solution 4.3.1.

(1) Let $a, b, x \in G$ and suppose $a * x = b * x$. If x^{-1} is the inverse of x, then

$$a * x = b * x \Longrightarrow a * \underbrace{x * x^{-1}}_{e} = b * \underbrace{x * x^{-1}}_{e} \Longrightarrow a * e = b * e.$$

Hence, $a = b$.

(2) Mutatis mutandis, readers may comfortably establish this statement.

Exercise 4.3.2. Let $(G, *)$ be a group and let $a, b, c \in G$. Assume that $a * b * c = e$ where e is the identity element of G.

(1) Does it follow that $b * c * a = e$?
(2) Does it follow that $b * a * c = e$?

Solution 4.3.2.

(1) True. Indeed,

$a * b * c = e$

$\Longrightarrow a^{-1} * a * b * c = a^{-1} * e$ (left multiplication by the inverse of a)

$\Longrightarrow e * b * c = a^{-1}$ (e is the identity element and a^{-1} is the inverse of a)

$\Longrightarrow b * c = a^{-1}$ (e is the identity element)

$\Longrightarrow b * c * a = a^{-1} * a$ (right multiplication by a)

$\Longrightarrow b * c * a = e$ (a^{-1} is the inverse of a).

(2) Untrue. On the group $(GL_2(\mathbb{R}), \cdot)$, take

$$A = \begin{pmatrix} 1 & 0 \\ 0 & -1 \end{pmatrix}, \ B = \begin{pmatrix} 0 & 2 \\ 1 & 0 \end{pmatrix} \text{ and } C = \begin{pmatrix} 0 & -1 \\ 1/2 & 0 \end{pmatrix}.$$

Then

$$AB = \begin{pmatrix} 0 & 2 \\ -1 & 0 \end{pmatrix}, \text{ and so } ABC = \begin{pmatrix} 1 & 0 \\ 0 & 1 \end{pmatrix} = I_2.$$

Nonetheless,

$$BA = \begin{pmatrix} 0 & -2 \\ 1 & 0 \end{pmatrix}, \text{ hence } BAC = \begin{pmatrix} -1 & 0 \\ 0 & -1 \end{pmatrix} \neq I_2,$$

as needed.

Exercise 4.3.3. Let $(G, *)$ be a group. An element x of G is called idempotent when $x * x = x$.

(1) Show that any group has exactly one idempotent element.
(2) Infer that none of $(\mathbb{Z}, \times)$, $(\mathbb{Q}, \times)$, $(\mathbb{R}, \times)$ and $(\mathbb{C}, \times)$ is a group.

Solution 4.3.3.

(1) Let $(G, *)$ be a group. Then G has an identity element e. By definition, $e * e = e$, and so e is idempotent. Let us show that there cannot be any other idempotent element. Let f be an element of G such that $f * f = f$. Then

$$f * f * f^{-1} = f * f^{-1} \text{ or } f * e = e.$$

As e is the identity element, $f * e = f$. Therefore, $f = e$, and this shows the uniqueness of the idempotent element in a group.

(2) Let G denote any of $\mathbb{Z}$, $\mathbb{Q}$, $\mathbb{R}$ or $\mathbb{C}$. Then

$$0 \times 0 = 0 \text{ and } 1 \times 1 = 1.$$

This says that 0 and 1 are two idempotent elements in G, and so $(G, \times)$ cannot be a group by the result of the preceding question.

Remark. This is an interesting criterion for testing whether a set equipped with a binary operation is a group. In case it is used to show that a certain set is not a group, its weakness lies in the fact that it does not say which axiom is violated.

Exercise 4.3.4. Let $G = \{x, y\}$ *(with $x \neq y$!)*. Define the binary operation $*$ as follows:

$$x * x = x * y = x, \ y * x = y * y = y.$$

Is $(G, *)$ a group?

Solution 4.3.4. First, a group always contains an identity element. In this case, if the identity element exists, then it is either x or y. Two cases are to be checked.

(1) If x is the identity element, then

$$x * x = x * x = x \text{ and } x * y = y * x = y.$$

By taking into account our assumptions above, we would then get $x = y$! Thus, x cannot be an identity element.

(2) The same reasoning allows us to conclude that y is not the identity element either! Therefore $(G, *)$ is not a group.

Exercise 4.3.5. Let $G = \{x, y\}$. Define $*$ as:

$$x * y = y * x = x, \; x * x = y * y = y.$$

Is $(G, *)$ a group?

Solution 4.3.5. In this exercise, $(G, *)$ is a group, even commutative!

(1) We start by checking that the binary operation $*$ satisfies the closure axiom. For all $x, y \in G$

$$x * y = y * x = x \in G \text{ and } x * x = y * y = y \in G.$$

(2) Readers will easily check that $*$ is associative by treating all possible cases.

(3) Since by hypothesis

$$x * y = y * x = x \text{ and } y * y = y,$$

we clearly see that y is the identity element for $*$.

(4) Each element of G admits an inverse. Indeed,

$$x * x = y \Longrightarrow x^{-1} = x$$

and

$$y * y = y \Longrightarrow y^{-1} = y.$$

(5) Commutativity was not part of the question, but notice that "$*$" is commutative.

Exercise 4.3.6. Let $\mathbb{Q}_+^*$ be the set of strictly positive rational numbers, then define an operation "$*$" on $\mathbb{Q}_+^*$ by:

$$a * b = \frac{a}{b}.$$

Is $(\mathbb{Q}_+^*, *)$ a group?

Solution 4.3.6. The binary operation $*$ fails to be associative, e.g. when $a = 1$, $b = 2$ and $c = 1/2$. Indeed, $a * b = 1/2$ and $b * c = 4$, thereby

$$(a * b) * c = 1 \neq 1/4 = a * (b * c).$$

Exercise 4.3.7. Let X be a non-empty set, and let $G = \mathcal{P}(X)$ be its powerset. Set

$$\forall A, B \in G : A * B = A \cup B.$$

Is $(G, *)$ a group?

Solution 4.3.7. First, recall that the elements of G are sets.

(1) Since $A \cup B$ is always a subset of $\mathcal{P}(X)$ (for all A and B) we see that $A * B \in G$, that is, $*$ satisfies the closure axiom.

(2) The operation $*$ is associative because if $A, B, C \in G$, then it is readily checked that

$$A * (B * C) = A \cup (B \cup C) = (A \cup B) \cup C = (A * B) * C.$$

(3) The identity element is $\varnothing$ since

$$\forall A \in G : \quad A \cup \varnothing = \varnothing \cup A = A.$$

(4) Not all elements have an inverse. Indeed, let $\varnothing \neq A \in G$. Then there is no $A' \in G$ such that

$$A * A' \ (= A \cup A') = A' * A = \ (A' \cup A) = \varnothing.$$

The reason is simple: For example, $A \cup A'$ contains always A and so it is never empty!

To conclude, $(G, *)$ is not a group.

Exercise 4.3.8. Let $G = \{1, 0, -1\}$. Is $(G, +)$ an abelian group, where "+" denotes the usual addition of numbers?

Solution 4.3.8. This is untrue. In fact G is not even closed under "+". Indeed, if this were true, then we would have $x + y \in G$ for all $x, y \in G$. But if $x = y = 1 \in G$, we see that

$$x + y = 1 + 1 = 2 \notin G,$$

and this violates the closure law.

Exercise 4.3.9. Is $(G, \times)$ a group in the following cases:

(1) $G = \{-1, 1\}$,
(2) $G = \{-1, 1, \frac{1}{3}, 3\}$,
(3) $G = \{2^n : n \in \mathbb{Z}\}$?

Solution 4.3.9.

(1) The operation "$\times$" obviously satisfies the closure axiom, and it is commutative and associative. The identity element is 1. The inverse of 1 is 1, and the inverse of -1 is -1. Thus, $(G, \times)$ is a commutative group.

(2) The operation "$\times$" does not satisfy the closure axiom because

$$-1 \times \frac{1}{3} = -\frac{1}{3} \notin G.$$

Thus, $(G, \times)$ is not a group.

(3) The operation "×" satisfies the closure axiom. Indeed,

$$\forall n, m \in \mathbb{Z} : 2^n 2^m = 2^{n+m} \in G.$$

It is commutative and associative. The identity element is $1 = 2^0$. The inverse of 2^n is 2^{-n} which belongs to G since $-n \in \mathbb{Z}$. Therefore, $(G, \times)$ is an abelian group.

Remark. See Exercise 4.3.34 for a more general case.

Exercise 4.3.10. Let $n \in \mathbb{N} - \{1\}$. Recall that the set

$$\mathbb{Z}_n = \mathbb{Z}/n\mathbb{Z} = \{\bar{0}, \bar{1}, \cdots, \overline{n-1}\}$$

represents all the equivalence classes under the relation $\mathcal{R}$ defined on $\mathbb{Z}$ by $x\mathcal{R}y$ iff $x - y$ is a multiple of n. Now, define on $\mathbb{Z}_n$ a binary operation $\oplus$ as

$$\bar{x} \oplus \bar{y} = \overline{x + y},$$

where $x, y \in \mathbb{Z}$ and where the sign $+$ in the expression $\overline{x + y}$ is the usual addition of numbers. Show that $(\mathbb{Z}/n\mathbb{Z}, \oplus)$ is an abelian group.

Solution 4.3.10. We must show that this new addition is well-defined. So, let $\bar{x} = \bar{x'}$ and $\bar{y} = \bar{y'}$. Then $x - x'$ and $y - y'$ are both multiples of n, hence so is their sum. That is,

$$x - x' + y - y' = x + y - (x' + y') = nk$$

for a certain integer k. Put differently, $\overline{x + y} = \overline{x' + y'}$, and so the choices of x and y are irrelevant. As for the closure law, just observe that if $x, y \in \mathbb{Z}$, then $x + y$ is clearly in one of the classes, noted $\bar{a}$, in $\mathbb{Z}_n$ for some n. In other words, $\bar{x} \oplus \bar{y} = \overline{x + y} = \bar{a} \in \mathbb{Z}_n$. So, "$\oplus$" is a well-defined binary operation.

To see why "$\oplus$" is commutative, let $\bar{x}, \bar{y} \in \mathbb{Z}_n$. Then

$$\bar{x} \oplus \bar{y} = \overline{x + y} = \overline{y + x} = \bar{y} \oplus \bar{x}$$

where, in the middle equality, we have switched the orders of x and y due to the commutativity of the usual addition on $\mathbb{Z}$.

As regards the associativity of "$\oplus$", it will be reduced to the associativity of "$+$" over $\mathbb{Z}$. This is left to readers.

The identity element is $\bar{0}$ as for all $\bar{x} \in \mathbb{Z}_n$

$$\bar{x} + \bar{0} = \overline{x + 0} = \bar{x}$$

since 0 is the identity element with respect to "$+$".

In the end, the inverse of each $\overline{x}$ is $\overline{-x}$ because

$$\overline{x} + \overline{-x} = \overline{x + (-x)} = \overline{0}.$$

Thus, $(\mathbb{Z}/n\mathbb{Z}, \oplus)$ is an abelian group.

Exercise 4.3.11. Let $n \in \mathbb{N} - \{1\}$. On $\mathbb{Z}_n$, define an operation $\otimes$ by

$$\overline{x} \otimes \overline{y} = \overline{x \cdot y},$$

where $x, y \in \mathbb{Z}$ and where the sign "$\cdot$" stands for the usual multiplication of numbers. Is $(\mathbb{Z}/n\mathbb{Z}, \otimes)$ an abelian group?

Solution 4.3.11. The answer is negative, i.e. $(\mathbb{Z}/n\mathbb{Z}, \otimes)$ is not an abelian group. We could pick the property which is not satisfied, and we would be done! However, and in this particular exercise, we shall also indicate those axioms which are satisfied. Besides their own interest, these will be used for some subsequent exercises.

The way of showing the closure law for "$\otimes$" is similar to that of the case "$\oplus$" treated above, a bit longer though. So, let $\overline{x} = \overline{x'}$ and $\overline{y} = \overline{y'}$. Then $x - x'$ and $y - y'$ are both multiples of n, hence so is their product. That is,

$$(x - x')(y - y') = nk$$

or $xy - x'y - xy' + x'y' = nk$. Hence

$$xy - x'y' = nk + x'y + xy' - 2x'y' = nk + x'(y - y') + (x - x')y'$$

is clearly a multiple of n. This leads to $\overline{x \cdot y} = \overline{x' \cdot y'}$, and so the operation is defined unambiguously. In the end, observe that if $x, y \in \mathbb{Z}$, then $x \cdot y$ is clearly in one of the classes in $\mathbb{Z}_n$ for some n. Thus, $\overline{x} \otimes \overline{y} = \overline{x \cdot y} \in \mathbb{Z}_n$. This settles the question of the "well-definedness" of the binary operation "$\otimes$".

To see why "$\otimes$" is associative, let $\overline{x}, \overline{y}, \overline{z} \in \mathbb{Z}_n$. Thanks to the associativity of the usual "$\cdot$" over $\mathbb{Z}$, one obtains

$$(\overline{x} \otimes \overline{y}) \otimes \overline{z} = \overline{(x \cdot y) \cdot z} = \overline{x \cdot (y \cdot z)} = \overline{x} \otimes (\overline{y} \otimes \overline{z}),$$

therefore showing that "$\otimes$" is associative.

By the commutativity of "$\cdot$" on $\mathbb{Z}$, it is seen that "$\otimes$" too is commutative on $\mathbb{Z}_n$, and details are left to readers.

Clearly, $\overline{1}$ is the identity element. Indeed, for all $\overline{x} \in \mathbb{Z}_n$

$$\overline{x} \otimes \overline{1} = \overline{x \cdot 1} = \overline{x}.$$

Nevertheless, not all elements of $\mathbb{Z}_n$ are invertible (witness $\overline{0}$). Indeed, if $\overline{0}$ were invertible, we would have for all $\overline{x}$

$$\overline{0} \otimes \overline{x} = \overline{0 \cdot x} = \overline{0} = \overline{1}$$

which is impossible.

Exercise 4.3.12. Set $\mathbb{Z}_n^* = \mathbb{Z}_n \setminus \{\overline{0}\}$. Is $(\mathbb{Z}_n^*, \otimes)$ a commutative group? If that is untrue, then under what condition on n, $(\mathbb{Z}_n^*, \otimes)$ becomes a (commutative) group?

Solution 4.3.12. While all laws satisfied by $\mathbb{Z}_n$ in the previous exercise remain valid in the case of $\mathbb{Z}_n^*$, we still have issues with inverses. Actually, we now have a much more serious problem. For instance, when $n = 4$ (which is not a random choice as shall be seen shortly), then $\mathbb{Z}_4^* = \{\overline{1}, \overline{2}, \overline{3}\}$. So, $\mathbb{Z}_4^*$ is not closed anymore under the law "$\otimes$". To see this, observe that $\overline{2} \in \mathbb{Z}_4^*$, but

$$\overline{2} \otimes \overline{2} = \overline{2 \cdot 2} = \overline{4} = \overline{0} \notin \mathbb{Z}_4^*.$$

On the other hand, readers may comfortably check that $(\mathbb{Z}_3^*, \otimes)$ is an abelian group. So, does this have something to do with odd and even numbers? Not quite, as this is rather related to prime numbers. Indeed: $(\mathbb{Z}_n^*, \otimes)$ is a (commutative) group if and only if n is a prime number. Let us show this fundamental result.

If n is not a prime number, writing $n = pq$ where $2 \leq p, q \leq n - 1$, then

$$\overline{p} \otimes \overline{q} = \overline{p \cdot q} = \overline{n} = \overline{0} \notin \mathbb{Z}_n^*.$$

If n is prime, let $\overline{a} \in \mathbb{Z}_n^*$. Clearly, $\gcd(a, n) = 1$. By Bézout's theorem in arithmetic, there are integers x and y such that $ax + ny = 1$. Using the notation of this exercise as well as that in Exercise 4.3.10, we obtain

$$\overline{1} = \overline{a \cdot x + n \cdot y} = \overline{a \cdot x} \oplus \overline{n \cdot y} = \overline{a \cdot x} = \overline{a} \otimes \overline{x}$$

as also $\overline{n \cdot y} = \overline{0}$. Since "$\otimes$" is commutative, it is seen that each $\overline{a}$ has an inverse. Therefore, $(\mathbb{Z}_n^*, \otimes)$ is a commutative group as long as n is a prime number.

Exercise 4.3.13. Let $\mathcal{F}(\mathbb{R})$ be the set of all functions from $\mathbb{R}$ into $\mathbb{R}$. Define three binary operation on $\mathcal{F}(\mathbb{R})$:

- Let $\circ$ be the usual composition of functions.
- Let $+$ be the usual addition of functions defined by

$$(\underbrace{f+g}_{\text{in } \mathcal{F}(\mathbb{R})})(x) = \underbrace{f(x) + g(x)}_{(+ \text{ in } \mathbb{R})}, \ \forall x \in \mathbb{R},$$

  and for all f and g in $\mathcal{F}$.
- Let $\cdot$ denote the pointwise product of functions, that is,

$$(\underbrace{f \cdot g}_{\text{in } \mathcal{F}(\mathbb{R})})(x) = \underbrace{f(x)g(x)}_{(\cdot \text{ in } \mathbb{R})}, \ \forall x \in \mathbb{R},$$

  and for all f and g in $\mathcal{F}(\mathbb{R})$.

Let $\mathcal{B}$ be the subset of $\mathcal{F}(\mathbb{R})$ constituted only of bijective functions, and let $\mathcal{F}(\mathbb{R}) \setminus \{0\}$ be the subset of $\mathcal{F}(\mathbb{R})$ constituted of functions f such that $f(x) \neq 0$ for each x.

(1) Is $(\mathcal{F}(\mathbb{R}), \circ)$ an abelian group?
(2) Is $(\mathcal{B}, \circ)$ an abelian group?
(3) Is $(\mathcal{F}(\mathbb{R}), +)$ an abelian group?
(4) Is $(\mathcal{F}(\mathbb{R}), \cdot)$ an abelian group?
(5) Is $(\mathcal{F}(\mathbb{R}) \setminus \{0\}, \cdot)$ an abelian group?

Solution 4.3.13.

(1) $(\mathcal{F}(\mathbb{R}), \circ)$ is not an abelian group: The closure law is clearly satisfied. The binary operation $\circ$ is always associative: Indeed, for all $f, g, h \in \mathcal{F}(\mathbb{R})$, one may write:

$$((f \circ g) \circ h)(x) = (f \circ g)(h(x)) = f(g(h(x))) = f(g \circ h(x)) = (f \circ (g \circ h))(x)$$

for all reals x. It is also easy to see that for all f

$$f \circ \mathrm{id} = \mathrm{id} \circ f = f$$

where id is the identity function on $\mathbb{R}$, i.e. $\mathrm{id}(x) = x$ for all $x \in \mathbb{R}$. So, the identity element is "id".

However, not every element in $\mathcal{F}(\mathbb{R})$ is invertible. Recall that a function f is invertible if and only if it is bijective. So, to find an element without an inverse, just consider any non-bijective function, e.g. $f(x) = x^2$ defined from $\mathbb{R}$ into $\mathbb{R}$.

In the end, observe that the composition of functions "$\circ$" is not commutative either. For instance, take $f(x) = x - 1$ and $g(x) = 2x$, both defined from $\mathbb{R}$ into $\mathbb{R}$. Then

$$\backsim (f \circ g)(x) = f(g(x)) = f(2x) = 2x - 1$$

whereas

$$(g \circ f)(x) = g(f(x)) = g(x - 1) = 2x - 2.$$

That is, $f \circ g \neq g \circ f$.

(2) Here, $(\mathcal{B}, \circ)$ is a group, still non-commutative though (the same counterexample just above works here as f and g are clearly bijective). Let us now show that $(\mathcal{B}, \circ)$ is a group. As regards the associative law, there is nothing to show for the binary operation is already associative in a larger set, namely $\mathcal{F}(\mathbb{R})$. The identity element is still "id". If $f \in \mathcal{B}$, then f is bijective and so by definition it has an inverse $f^{-1} \in \mathcal{B}$. Thus, we have shown that $(\mathcal{B}, \circ)$ is a group.

Remark. If B denotes now the set of all bijective functions from a set E onto E, then it may be shown that $(\mathcal{B}, \circ)$ is still a group.

(3) $(\mathcal{F}(\mathbb{R}), +)$ is an abelian group. This follows basically from the commutativity of the group $(\mathbb{R}, +)$. Indeed, the binary operation "$+$" is associative: Let $f, g, h \in \mathcal{F}(\mathbb{R})$. Then for all $x \in \mathbb{R}$:

$$(f(x) + g(x)) + h(x) = f(x) + g(x) + h(x) = f(x) + (g(x) + h(x))$$

where we have used the associativity of "$+$" over $\mathbb{R}$ because $f(x), g(x), h(x) \in \mathbb{R}$ for all $x \in \mathbb{R}$.

Since for all $f \in \mathcal{F}(\mathbb{R})$, $f + 0 = 0 + f = f$ (where 0 designates here the zero function) because

$$f(x) + 0 = 0 + f(x) = f(x)$$

for all $x \in \mathbb{R}$ as $0 \in \mathbb{R}$ is the identity element of $(\mathbb{R}, +)$, we see that the identity element of $(\mathcal{F}(\mathbb{R}), +)$ is the zero function, i.e. $e(x) = 0$ for all $x \in \mathbb{R}$. Readers may check that each $f \in \mathcal{F}(\mathbb{R})$ has an inverse given by $-f$ which clearly an element of $\mathcal{F}(\mathbb{R})$. Finally, as $f + g = g + f$ for all $f, g \in \mathcal{F}(\mathbb{R})$ since $f(x) + g(x) = g(x) + f(x)$ for any $x \in \mathbb{R}$ (which is true as $f(x), g(x) \in \mathbb{R}$ and the addition of real numbers is commutative). Consequently, $(\mathcal{F}(\mathbb{R}), +)$ is an abelian group.

(4) $(\mathcal{F}(\mathbb{R}), \cdot)$ is not an abelian group. The binary operation "$\cdot$" satisfies the closure, associative, commutative, and identity laws (the identity being the function $x \mapsto f(x) = 1$). However, not all functions have an inverse w.r.t. "$\cdot$" (for example, the zero function does not possess any inverse in $(\mathcal{F}(\mathbb{R}), \cdot)$).

(5) This time, $(\mathcal{F}(\mathbb{R})\backslash\{0\}, \cdot)$ is indeed an abelian group. The same reasoning as in the foregoing answer applies, but now every function f possesses an inverse given by $x \mapsto f^{-1}(x) = 1/f(x)$.

Exercise 4.3.14. Let G be the set of all increasing functions from $\mathbb{R}$ into $\mathbb{R}$. Is $(G, +)$ a group?

Solution 4.3.14. Recall that an increasing function $f : \mathbb{R} \to \mathbb{R}$ is one which satisfies $f(x) \leq f(y)$ whenever $x \leq y$. Observe that $(G, +)$ satisfies all laws of a group except the inverse law. Indeed, the function $x \mapsto x$ defined from $\mathbb{R}$ into $\mathbb{R}$ is increasing, hence it belongs to G, whilst the "possible inverse would be" $h(x) = -x$ defined from $\mathbb{R}$ into $\mathbb{R}$. Since $h \notin G$ as h is non-increasing (not because it is decreasing!), it follows that $(G, +)$ is not a group.

Exercise 4.3.15. Let $G = \{f, g, h, i\}$ where f, g, h and k are functions defined from $\mathbb{R}^*$ into $\mathbb{R}^*$ by

$$f(x) = x, \ g(x) = \frac{1}{x}, \ h(x) = -x \text{ and } k(x) = -\frac{1}{x}$$

respectively. Let "$\circ$" be the usual function composition.
  (1) Prove that $(G, \circ)$ is a group.
  (2) Is it commutative?

Solution 4.3.15.
  (1) (a) The binary operation $\circ$ satisfies the closure axiom because any of the possible compositions gives one of the following cases
$$x, \ \frac{1}{x}, \ -x, \ -\frac{1}{x},$$
    i.e. each time we find an element of G.
  (b) The operation $\circ$ is associative (this is left to readers).
  (c) Let e be the identity element (if it exists). Then
$$e \circ l = l \circ e = l, \ \forall l \in G.$$

We can see that:
$$f \circ g = g \circ f = g,$$
$$f \circ h = h \circ f = h,$$
$$f \circ k = k \circ f = k$$
and
$$f \circ f = f \circ f = f.$$
So, $f \ (\in G)$ is the sought identity element.

(d) We have:

$$k \circ k = k \circ k = f,$$
$$g \circ g = g \circ g = f$$
$$h \circ h = h \circ h = f$$

and

$$f \circ f = f \circ f = f$$

(observe that the inverse of each element of G is itself). Thus, $(G, \circ)$ is a group.

(2) Readers may easily verify that the operation $\circ$ is commutative.

Exercise 4.3.16. Show that $(M_2(\mathbb{R}), +)$ is an abelian group.

Solution 4.3.16. By the definition of the addition of matrices, we see that $M_2(\mathbb{R})$ is closed under addition. It is also easy to see that "$+$" is associative. Indeed, for $A = \begin{pmatrix} a & b \\ c & d \end{pmatrix}$, $A' = \begin{pmatrix} a' & b' \\ c' & d' \end{pmatrix}$ and $A'' = \begin{pmatrix} a'' & b'' \\ c'' & d'' \end{pmatrix}$ in $M_2(\mathbb{R})$:

$$
\begin{aligned}
(A + A') + A'' &= \left[\begin{pmatrix} a & b \\ c & d \end{pmatrix} + \begin{pmatrix} a' & b' \\ c' & d' \end{pmatrix} \right] + \begin{pmatrix} a'' & b'' \\ c'' & d'' \end{pmatrix} \\
&= \begin{pmatrix} a + a' & b + b' \\ c + c' & d + d' \end{pmatrix} + \begin{pmatrix} a'' & b'' \\ c'' & d'' \end{pmatrix} \\
&= \begin{pmatrix} (a + a') + a'' & (b + b') + b'' \\ (c + c') + c'' & (d + d') + d'' \end{pmatrix} \\
&= \begin{pmatrix} a + (a' + a''). & b + (b' + b'') \\ c + (c' + c'') & d + (d' + d'') \end{pmatrix} \text{ (as + is associative in } \mathbb{R}) \\
&= \begin{pmatrix} a & b \\ c & d \end{pmatrix} + \begin{pmatrix} a' + a'' & b' + b'' \\ c' + c'' & d' + d'' \end{pmatrix} \\
&= \begin{pmatrix} a & b \\ c & d \end{pmatrix} + \left[\begin{pmatrix} a' & b' \\ c' & d' \end{pmatrix} + \begin{pmatrix} a'' & b'' \\ c'' & d'' \end{pmatrix} \right] \\
&= A + (A' + A'').
\end{aligned}
$$

The identity element with respect to "$+$" is $\begin{pmatrix} 0 & 0 \\ 0 & 0 \end{pmatrix}$ (called the zero matrix) because

$$\begin{pmatrix} a & b \\ c & d \end{pmatrix} + \begin{pmatrix} 0 & 0 \\ 0 & 0 \end{pmatrix} = \begin{pmatrix} 0 & 0 \\ 0 & 0 \end{pmatrix} + \begin{pmatrix} a & b \\ c & d \end{pmatrix} = \begin{pmatrix} a & b \\ c & d \end{pmatrix}$$

for any $A = \begin{pmatrix} a & b \\ c & d \end{pmatrix} \in M_2(\mathbb{R})$.

The inverse of each $\begin{pmatrix} a & b \\ c & d \end{pmatrix}$ is $\begin{pmatrix} -a & -b \\ -c & -d \end{pmatrix}$ (which is patently an element of $M_2(\mathbb{R})$) as

$$\begin{pmatrix} a & b \\ c & d \end{pmatrix} + \begin{pmatrix} -a & -b \\ -c & -d \end{pmatrix} = \begin{pmatrix} -a & -b \\ -c & -d \end{pmatrix} + \begin{pmatrix} a & b \\ c & d \end{pmatrix} = \begin{pmatrix} 0 & 0 \\ 0 & 0 \end{pmatrix}.$$

Finally, for any $A = \begin{pmatrix} a & b \\ c & d \end{pmatrix}$ and any $B = \begin{pmatrix} x & y \\ z & t \end{pmatrix}$, we have

$$A + B = \begin{pmatrix} a+x & b+y \\ c+z & d+t \end{pmatrix} = \begin{pmatrix} x+a & y+b \\ z+c & t+d \end{pmatrix} = B + A$$

(due to the commutativity of "+" in $\mathbb{R}$). Consequently, we have shown that $(M_2(\mathbb{R}), +)$ is an abelian group, as required.

Exercise 4.3.17. Over $\mathbb{R}$, we define the binary operation $*$ by

$$\forall(x, y) \in \mathbb{R}^2 : \ x * y = 4xy + x + y.$$

Is $(\mathbb{R}, *)$ a commutative group? If this is not the case, give the largest subset of $\mathbb{R}$, noted A, such that $(A, *)$ becomes an abelian group.

Solution 4.3.17. Let us proceed as if we were asked to show that $(\mathbb{R}, *)$ is a group.

(1) First, we must check that the operation $*$ satisfies the closure axiom, i.e.

$$\forall x, y \in \mathbb{R} : \ x * y \in \mathbb{R}.$$

Let $x, y \in \mathbb{R}$. By the axioms of real numbers, $4xy + x + y \in \mathbb{R}$, i.e. $x * y \in \mathbb{R}$. Hence, $\mathbb{R}$ is closed under "$*$".

(2) The operation $*$ is associative. Let $x, y, z \in \mathbb{R}$. We have

$$\begin{aligned}
(x * y) * z &= 4(x * y)z + (x * y) + z \\
&= 4(4xy + x + y)z + 4xy + x + y + z \\
&= 16xyz + 4xy + 4xz + 4yz + x + y + z.
\end{aligned}$$

On the other hand,

$$\begin{aligned}
x * (y * z) &= 4x(y * z) + x + (y * z) \\
&= 4x(4yz + y + z) + x + (4yz + y + z) \\
&= 16xyz + 4xy + 4xz + 4yz + x + y + z.
\end{aligned}$$

Therefore

$$\forall x, y, z \in \mathbb{R} : \ (x * y) * z = x * (y * z),$$

which shows the associativity of "$*$".

(3) The operation $*$ admits an identity element. We have to find some $e \in \mathbb{R}$ such that $e * x = x * e = x$ for all $x \in \mathbb{R}$. Since

$$\forall x \in \mathbb{R} : \ x * 0 = 0 * x = x,$$

it is seen that 0 is the identity element with respect to "$*$".

(4) Does any x of $\mathbb{R}$ admit an inverse with respect to "$*$"? That is, do we have

$$\forall x \in \mathbb{R}, \ \exists x' \in \mathbb{R} : \ x * x' = x' * x = 0?$$

The answer is negative, and to see that, let $x \in \mathbb{R}$ be such that $x' * x = 0$. So

$$4x'x + x' + x = 4xx' + x + x' = 0 \iff 4xx' + x + x' = 0$$

(by the commutativity of the usual multiplication and addition in $\mathbb{R}$). We get

$$x' = \frac{-x}{4x + 1}, \quad \text{when } x \neq -\frac{1}{4}.$$

So, each $x \neq -1/4$ has an inverse.

Now, we show that $x = -\frac{1}{4}$ does not possess an inverse. If $x = -\frac{1}{4}$ had an inverse, we would have

$$-\frac{1}{4} * x' = 0 \text{ or } -x' + x' - \frac{1}{4} = 0$$

which gives $-\frac{1}{4} = 0$, a contradiction! So, $-\frac{1}{4}$ does not admit an inverse. The structure of a group is therefore violated.

In the end, the binary operation $*$ is clearly commutative for the addition and the multiplication in $\mathbb{R}$ are commutative.

To answer the second question, it is clear to readers that $A = \mathbb{R} \setminus \{-\frac{1}{4}\}$, and so $(\mathbb{R} \setminus \{-\frac{1}{4}\}, *)$ is an abelian group. Everything is as above, and the only thing which requires perhaps a little clarification is the closure law: We must show that for all $x, y \in \mathbb{R} \setminus \{-\frac{1}{4}\}$: $x * y \in \mathbb{R} \setminus \{-\frac{1}{4}\}$. We use a proof by contraposition. Assume that $x * y = -1/4$. Hence

$$4xy + x + y = -1/4 \implies 16xy + 4x + 4y + 1 = 0 \implies (4x + 1)(4y + 1) = 0$$

which gives either $x = -1/4$ or $y = -1/4$, as needed.

Exercise 4.3.18. On $G = (-1, 1)$, define $*$ by

$$\forall x, y \in G : \ x * y = \frac{x + y}{1 + xy}.$$

Show that $(G, *)$ is an abelian group.

Solution 4.3.18.

(1) Let's check that $*$ satisfies the closure axiom. Let $x, y \in G$. We ought to show that $x * y \in G$. One of the ways of doing that is to fall back on basic calculus. Define the function f on $[-1, 1]$ by
$$f(t) = \frac{t + y}{1 + ty}$$
(where $y \in (-1, 1)$ is fixed for now), and so $f(x) = x * y$.

We have
$$\forall t \in [-1, 1] : f'(t) = \frac{1 - y^2}{(1 + ty)^2} > 0.$$

Since $[-1, 1]$ is an interval and f is strictly increasing on $[-1, 1]$, we obtain:
$$f(-1) < f(x) < f(1).$$

But
$$f(x) = x * y, \; f(-1) = \frac{-1 + y}{1 - y} = -1 \text{ and } f(1) = \frac{1 + y}{1 + y} = 1.$$

In other words,
$$\forall x, y \in (-1, 1) : \; x * y \in (-1, 1).$$

(2) The binary operation $*$ is associative because for all $(x, y, z) \in G^3$, we have:
$$
\begin{aligned}
x * (y * z) &= \frac{x + (y * z)}{1 + x(y * z)} \\
&= \frac{x + \frac{y + z}{1 + yz}}{1 + x \frac{y + z}{1 + yz}} \\
&= \frac{x + y + z + xyz}{1 + xy + xz + yz}.
\end{aligned}
$$

Similarly, we find
$$(x * y) * z = \frac{x + y + z + xyz}{1 + xy + xz + yz},$$
whereby $*$ is associative.

(3) Let $x \in G$ be such that
$$x * e = e * x = x.$$

We have
$$x * e = x \iff \frac{x + e}{1 + xe} = x \iff x + e = x + x^2 e \iff (1 - x^2)e = 0,$$
and since $x \neq \pm 1$, we get $e = 0$.

In a similar way, we get from $e*x = x$ that $e = 0$. Therefore $e = 0$ is indeed the identity element because

$$\exists e = 0 \in (-1, 1), \forall x \in (-1, 1): \ x * e = e * x = x.$$

(4) Let $x \in G$, and consider the equations

$$x * x' = x' * x = e = 0,$$

where x' is the unknown. We have

$$x * x' = 0 \iff \frac{x + x'}{1 + xx'} = 0,$$

but since $1 + xx' \neq 0$ for all $(x, x') \in G^2$, it is seen that $x' = -x$. We now check a crucial point (which is pretty simple here) that students usually tend to forget. Do we have $x' \in G = (-1, 1)$? Yes, we do as $x \in (-1, 1)$ yields $x' = -x \in (-1, 1)$.

Similarly, we show that $x' = -x$ is also a solution of $x' * x = 0$. Thus,

$$\forall x \in (-1, 1), \exists x' = -x \in (-1, 1): \ x * x' = x' * x = 0,$$

i.e. each element of G admits an inverse. ·

To summarize, $(G, *)$ is indeed a group.

(5) This group is also commutative because for all x and y in G, we have (thanks to the commutativity of the addition and multiplication of the real numbers)

$$x * y = \frac{x + y}{1 + xy} = \frac{y + x}{1 + yx} = y * x.$$

Remark. Since we knew from the question that "$*$" was going to be commutative, we should have started by showing that the binary operation is commutative. Then, when it comes to the identity element (resp. the inverse element) we would need only treat one of the equations $x * e = x$ or $e * x = x$ (resp. $x * x' = e$ or $x' * x = e$). We have not done that here so that students get used to treat both equations as commutativity is not available all the time.

Remark. Using the same method, we can prove that $G = (-c, c)$ (with $c > 0$) with respect to the binary operation $*$ defined by

$$\forall x, y \in G: \ x * y = \frac{x + y}{1 + xy/c^2}$$

is a commutative group. P. J. Cameron noted in [7] that this example is used in special relativity to describe the addition of velocities.

Exercise 4.3.19. We define on $\mathbb{R}$, a binary operation $*$ by:

$$x * y = \begin{cases} \frac{x^3+y^3}{x^2+y^2}, & (x,y) \neq (0,0), \\ 0, & (x,y) = (0,0). \end{cases}$$

Prove that $*$ is commutative, admits an identity element and every real number has an inverse with respect to "$*$", and yet $*$ is not associative.

Solution 4.3.19.

(1) Since $(\mathbb{R}, +)$ is commutative, for all x and y (non-null simultaneously)

$$x * y = \frac{x^3 + y^3}{x^2 + y^2} = \frac{y^3 + x^3}{y^2 + x^2} = y * x$$

and

$$0 * 0 = 0 * 0 \, (= 0).$$

So, "$*$" is commutative.

(2) 0 is the identity element for

$$\forall x \neq 0 : \ x * 0 = \frac{x^3 + 0}{x^2 + 0} = \frac{x^3}{x^2} = x$$

and

$$0 * 0 = 0 * 0 = 0.$$

(3) For all x, $-x$ is its inverse since

$$x * (-x) = \frac{x^3 + (-x)^3}{x^2 + (-x)^2} = \frac{x^3 - x^3}{x^2 + (-x)^2} = \frac{0}{2x^2} = 0$$

if $x \neq 0$, and when $x = 0$, then $0 * 0 = 0$ (and so the inverse of 0 is 0).

(4) $*$ is not associative as

$$(-1) * \underbrace{[(1 * 1)]}_{1} = (-1) * 1 = 0$$

and

$$[(-1) * 1] * 1 = 0 * 1 = 1.$$

Hence,

$$(-1) * [(1 * 1)] \neq [(-1) * 1)] * 1.$$

Consequently, $(\mathbb{R}, *)$ is not a group.

We have another way of showing that "$*$" is not associative. Since $1 * 1 = 1$ and $0 * 0 = 0$, $(\mathbb{R}, *)$ is not a group by Exercise 4.3.3. Since the closure, identity and inverse laws are satisfied, it follows that "$*$" is not associative.

Exercise 4.3.20. Let $\mathbb{C}$ be the set of complex numbers and let

$$U = \{z \in \mathbb{C} : |z| = 1\}.$$

Show that $(U, \times)$ is a group, where "$\times$" denotes the usual multiplication.

Solution 4.3.20.

(1) Let $z, z' \in U$. First, we prove that $z \times z'$ belongs to U. Obviously, $z \times z'$ is a complex number and

$$|z \times z'| = |z| \times |z'| = 1 \times 1 = 1.$$

Hence, $z \times z' \in U$.

(2) Clearly, for all z, z' and z'' in U

$$z \times (z' \times z'') = (z \times z') \times z'',$$

(as this is already available in $(\mathbb{C}, \times)$), i.e. "$\times$" is associative.

(3) Let $z \in U$. The apparent solution (and independent of z) of $z \times e = z$ and $e \times z = z$ is $e = 1 \in U$, which is the identity element.

(4) Let $z \in U$. Let us find $z' \in U$ such that

$$z \times z' = 1 \text{ and } z' \times z = 1.$$

We get $z' = 1/z$, which is well-defined since $z \neq 0$, because $|z| = 1$. Moreover, it must be checked that z' belongs to U. This is definitely the case for

$$|z'| = \left|\frac{1}{z}\right| = \frac{1}{|z|} = \frac{1}{1} = 1.$$

Thus, $(U, \times)$ is a group, as suggested.

Exercise 4.3.21. On $\mathbb{R}^2$, define a binary operation, noted $\oplus$, by

$$(x, y) \oplus (x', y') = (x + x', y + y')$$

for all $(x, y), (x', y') \in \mathbb{R}^2$. Show that $(\mathbb{R}^2, \oplus)$ is an abelian group (it is customary to denote the operation $\oplus$ by $+$).

Solution 4.3.21.

(1) Let $(x, y), (x', y') \in \mathbb{R}^2$. Since $x + x, y + y' \in \mathbb{R}$, it is seen that $(x, y) \oplus (x', y') \in \mathbb{R}^2$, i.e. $\mathbb{R}^2$ is closed under the operation $\oplus$.

(2) Let $(x, y), (x', y') \in \mathbb{R}^2$. Since $+$ is commutative in $\mathbb{R}$, we may write

$$(x, y) \oplus (x', y') = (x + x', y + y') = (x' + x, y' + y) = (x', y') \oplus (x, y).$$

In other language, $\oplus$ is commutative.

(3) To see why $\oplus$ is associative, let $(x, y), (x', y'), (x'', y'') \in \mathbb{R}^2$.
Then, we have on the one hand

$$[(x, y) \oplus (x', y')] \oplus (x'', y'') = (x + x', y + y') \oplus (x'', y'') = (x + x' + x'', y + y' + y''),$$

and on the other hand,

$$(x, y) \oplus [(x', y') \oplus (x'', y'')] = (x, y) \oplus (x' + x'', y +' y'') = (x + x' + x'', y + y' + y'').$$

Thus, $\oplus$ is indeed associative.

(4) To find the identity element, noted (e, f), let $(x, y) \in \mathbb{R}^2$ be
such that

$$(x, y) = (x, y) \oplus (e, f) = (x + e, y + f)$$

(observe that there is no need to deal with the equation $(x, y) = (e, f) \oplus (x, y)$ thanks to the commutativity of $\oplus$). Therefore,
$(e, f) = (0, 0)$ is the identity element.

(5) Let $(x, y) \in \mathbb{R}^2$. Since $\oplus$ is commutative, we find (x', y') such
that, e.g.

$$(0, 0) = (x, y) \oplus (x', y') = (x + x', y + y'),$$

and so $x' = -x$ and $y' = -y$. Thence, each (x, y) has an
inverse given by $(-x, -y)$ (which is clearly an element of $\mathbb{R}^2$).

Accordingly, we have shown that $(\mathbb{R}^2, \oplus)$ is an abelian
group.

Exercise 4.3.22. On $\mathbb{R}^2$, define a binary operation, denoted by $*$,
by

$$(x, y) * (x', y') = (xx', yx' + y')$$

for all $(x, y), (x', y') \in \mathbb{R}^2$. Is $(\mathbb{R}^2, *)$ is an abelian group?

Solution 4.3.22. First, observe that $\mathbb{R}^2$ is closed under the binary
operation $*$.

(1) The operation "$*$" is associative. Let the interested reader do
the details.

(2) To find the identity element (e, f), we need to solve the system

$$(x, y) = (x, y) * (e, f) = (xe, ye + f),$$

i.e. $x = xe$ and $ye + f = y$, from which we derive $(e, f) = (1, 0)$.
Readers must also solve $(e, f) * (x, y) = (x, y)$ which also gives
$(e, f) = (1, 0)$ (especially that "$*$" is not commutative, as we
will shortly see).

(3) All elements of the type $(0, a)$ do not possess an inverse, and
this may easily be established by readers.

(4) In the end, "$*$" is not commutative as e.g.

$$(0,1) * (2,1) = (0,3) \neq (0,1) = (2,1) * (0,1).$$

To summarize, $(\mathbb{R}^2, *)$ is not a group, and "$*$" is not commutative.

Exercise 4.3.23. Let $(G, *)$ be a group and let e be its identity element. Assume that:

$$\forall x \in G: \ x^2 = e,$$

where $x^2 = x * x$.

(1) Show that $x^{-1} = x$.
(2) Deduce that G is abelian.

Solution 4.3.23.

(1) Let $x \in G$. We can write

$$x * x = e \iff \underbrace{x^{-1} * x}_{=e} *x = x^{-1} *e = x^{-1} \iff e*x = x^{-1} \iff x = x^{-1}.$$

(2) Let $x, y \in G$. Then $x * y \in G$ as G is a group. By the first question, we get

$$x = x^{-1}, \ y = y^{-1} \text{ and } (x * y)^{-1} = x * y.$$

Since we already know that

$$(x * y)^{-1} = y^{-1} * x^{-1},$$

it follows that

$$x * y = (x * y)^{-1} = y^{-1} * x^{-1} = y * x,$$

which is exactly what we have been after, i.e. the commutativity of "$*$".

Exercise 4.3.24. Let $(G, *)$ be a group where e is its identity element. Say that $x \in G$ is right invertible provided $x * a = e$ for some $a \in G$. If $b * x = e$ for a certain $b \in G$, then x is called left invertible.

(1) Show that if an x is both right and left invertible, then it is invertible.
(2) Infer that the inverse, when it exists, is unique.
(3) Find a left invertible element in a group which is not invertible, that is, it is not right invertible.

Solution 4.3.24.

(1) By assumption, $x * a = e$ for some $a \in G$, and $b * x = e$ for a certain $b \in G$. Hence

$$b = b * e = b * x * a = e * a = a.$$

Therefore, $x * a = a * x = e$, and so x is invertible.

(2) Assume x has two inverses, a and b say. Then, $a = b$ as before. So the inverse, when it exists, is unique.

(3) Let $X = \{0\}$ and $Y = \{0,1\}$, then define $f : X \to Y$ by $f(0) = 0$, and $g : Y \to X$ by $g(0) = g(1) = 0$. Now consider the group of functions from $\mathbb{R}$ into $\mathbb{R}$ equipped with the group operation "$\circ$". Recall that in this context, f has a left inverse if $g \circ f(x) = x$ for all $x \in X$, i.e. for $x = 0$ here only. This is in effect the case for

$$g \circ f(0) = g[f(0)] = g(0) = 0.$$

If f is right invertible, according to the first question, the right inverse must be the function g. However,

$$f \circ g(1) = f[g(1)] = f(0) = 0 \neq 1,$$

that is, f is not right invertible.

Exercise 4.3.25. (Cf. [5]) Consider the function $f : \mathbb{N} \to \mathbb{N}$ defined by $f(n) = n^2$. Show that f has no right inverse, and exhibit two left inverses of f.

Solution 4.3.25. Assume for the sake of contradiction that f has a right inverse $g : \mathbb{N} \to \mathbb{N}$. Then $(f \circ g)(n) = n$ for all $n \in \mathbb{N}$, i.e.

$$f[g(n)] = [g(n)]^2 = n$$

for all n. Since $g(n) \in \mathbb{N}$, it is seen that the previous equation is not true for all n (witness $n = 2$).

Let us exhibit two left inverses of f. Set

$$g_1(n) = \begin{cases} \sqrt{n}, & \text{when } n \text{ is a perfect square,} \\ n+1, & \text{otherwise,} \end{cases}$$

and

$$g_2(n) = \begin{cases} \sqrt{n}, & \text{when } n \text{ is a perfect square,} \\ n+2, & \text{otherwise,} \end{cases}$$

both defined from $\mathbb{N}$ into $\mathbb{N}$. Then, for all $n \in \mathbb{N}$

$$(g_1 \circ f)(n) = g_1[f(n)] = g_1(n^2) = n,$$

and similarly $(g_2 \circ f)(n) = n$. This shows that g_1 and g_2 are two left inverses of f.

Remark. Using the same idea, show that f has infinitely many left inverses.

Exercise 4.3.26. On $\mathbb{R}^2$, define a binary operation, noted $*$, by

$$(x, y) * (x', y') = (xx' + yy', xy' + yx')$$

for all $(x, y), (x', y') \in \mathbb{R}^2$. Find

$$(x, y)^n = (x, y) * (x, y) * \cdots * (x, y)$$

(n times).

Solution 4.3.26. Readers may check using induction on n that

$$(x, y)^n = \left(\frac{(x + y)^n + (x - y)^n}{2}, \frac{(x + y)^n - (x - y)^n}{2} \right).$$

Exercise 4.3.27. Let $(G, *)$ be an abelian group, with e being the identity element. Show that for all $x, y \in G$ and all integers n:

$$(x * y)^n = x^n * y^n.$$

Solution 4.3.27. First, we treat the case of positive integers n, and we use a proof by induction on n. The statement obviously holds for $n = 0$ (both sides equal e). Suppose $(x * y)^n = x^n * y^n$, and we show that $(x * y)^{n+1} = x^{n+1} * y^{n+1}$. We have

$$(x * y)^{n+1} = (x * y)^n * (x * y) = x^n * y^n * x * y = x^n * x * y^n * y$$

(where we have used the induction hypothesis as well as the commutativity of $*$). Therefore,

$$(x * y)^{n+1} = x^{n+1} * y^{n+1},$$

and this settles the case of positive integers.

If n is a negative integer, $-n$ is a positive integer, call it m. By the first part of the proof, we know that

$$(x * y)^m = x^m * y^m.$$

Hence

$$(x * y)^n = (x * y)^{-m} = [(x * y)^m]^{-1} = (x^m * y^m)^{-1} = (y^m)^{-1} * (x^m)^{-1} = y^{-m} * x^{-m}.$$

Because $(G, *)$ is an abelian group, and $n = -m$, we obtain

$$(x * y)^n = x^n * y^n$$

for negative n. The proof is therefore complete.

Exercise 4.3.28. If $(G, *)$ is a group such that for all $x, y \in G$: $(x * y)^2 = x^2 * y^2$, show that $(G, *)$ must be abelian.

Solution 4.3.28. Let $x, y \in G$ be such that $(x * y)^2 = x^2 * y^2$, and let us show that "$*$" is commutative. Clearly,

$$(x * y)^2 = x^2 * y^2 \iff x * y * x * y = x * x * y * y.$$

By using each of the two properties of Exercise 4.3.1, we get $x*y = y*x$, i.e. we have shown the commutativity of "$*$".

Exercise 4.3.29. Let $(G, *)$ be a group such that card $G = 3$. Find the Cayley table.

Solution 4.3.29. Let $G = \{e, x, y\}$ where e is the identity element. The table to be filled in is:

$*$	e	x	y
e			
x			
y			

By definition of the identity element, we have

$*$	e	x	y
e	e	x	y
x	x		
y	y		

Now, if $x * x = e$, then we must have $x * y = y$. But

$$x * y = e * y$$

and by simplification (see Exercise 4.3.1), $x = e$. So $x * x = y$ and hence $x * y = e$.

Therefore, the required table is:

$*$	e	x	y
e	e	x	y
x	x	y	e
y	y	e	x

Exercise 4.3.30. Show that $3\mathbb{Z}$ is a subgroup of $(\mathbb{Z}, +)$ where $3\mathbb{Z}$ denotes the multiples of 3.

Solution 4.3.30. First, $3\mathbb{Z}$ is non-empty. Next, let x, y be two elements of $3\mathbb{Z}$, so they can be written as $3p$ and $3q$ respectively, where $p, q \in \mathbb{Z}$. The inverse of y is $-y$, and so $-y = -3q$. Then we have

$$x - y = x + (-y) = 3p + (-3q) = 3p - 3q = 3(p - q) \in 3\mathbb{Z}$$

because $p - q \in \mathbb{Z}$. As a result, $(3\mathbb{Z}, +)$ is a subgroup of $(\mathbb{Z}, +)$.

Remark. There is nothing special about the number 3 in this exercise. In fact, it can be shown in very much the same way that the subgroups of $(\mathbb{Z}, +)$ are $n\mathbb{Z}$, $n \in \mathbb{N}$.

Exercise 4.3.31. Determine in each case whether the given set H defines a subgroup of G (denote the identity element of G by e_G):

(1) $H = \mathbb{Q}$, $G = (\mathbb{Z}, +)$.
(2) $H = (0, \infty)$, $G = (\mathbb{R}, +)$.
(3) $H = [-1, 1]$, $G = (\mathbb{R}, +)$.
(4) $H = \mathbb{Q}^*$, $G = (\mathbb{R}, \times)$.
(5) $H = \mathbb{Q}^*$, $G = (\mathbb{R}^*, \times)$.
(6) $H = [1, \infty)$, $G = (\mathbb{R}^*, \times)$.
(7) $H = (0, \infty)$, $G = (\mathbb{R}^*, \times)$.

(8) H is the set of 2×2 real matrices of the form $\begin{pmatrix} 0 & a \\ b & 1 \end{pmatrix}$ with $a, b \in \mathbb{R}$, $G = (GL_2(\mathbb{R}), \cdot)$.

(9) H is the set of 2×2 real matrices of the form $\begin{pmatrix} 1/a & 0 \\ 0 & a \end{pmatrix}$ with $a \neq 0$, $G = (GL_2(\mathbb{R}), \cdot)$.

Solution 4.3.31.

(1) H is not even a subset of G!
(2) H is not a subgroup of G because $e_G = 0 \notin H$.
(3) H (which is non-empty) is a subgroup of G in this case signifies that

$$\forall x, y \in H : x + (-y) = x - y \in H.$$

So, H is not a subgroup of G because, e.g. $1, -1/2 \in H$, but $1 - (-1/2) = 3/2 \notin H$.

(4) G is not a group!
(5) In this case, H is a subgroup of G. Let $r, s \in \mathbb{Q}^*$. We must show that $rs^{-1} = r/s \in \mathbb{Q}^*$. Since $r, s \in \mathbb{Q}^*$, we know that $r = a/b$ and $s = c/d$, where a, b, c, d are all non-zero integers. So

$$rs^{-1} = \frac{r}{s} = \frac{ad}{bc} \in \mathbb{Q}^*,$$

as needed.

(6) H is not a subgroup of G. For example, the "inverse" of $2 \in H$ would be $1/2 \notin H$.
(7) We show that H is a subgroup of G (obviously, $e_G = 1 \in H$). Besides, if $x, y \in H$, i.e. $x, y > 0$, then $xy^{-1} > 0$, that is, $xy^{-1} \in H$, as needed.

(8) H is not a subgroup for the mere reason that e_G, i.e. the matrix $\begin{pmatrix} 1 & 0 \\ 0 & 1 \end{pmatrix}$, does not belong to H.

(9) In this case, H is a subgroup of G. First, $H \subset GL_2(\mathbb{R})$ for the determinant of each element of H is equal to 1. It is also plain that $H \neq \varnothing$. We know from Proposition 4.1.8 that the inverse of an element of H is of the form $\begin{pmatrix} a & 0 \\ 0 & 1/a \end{pmatrix}$, with $a \neq 0$. So, let $A = \begin{pmatrix} 1/a & 0 \\ 0 & a \end{pmatrix}, B = \begin{pmatrix} 1/b & 0 \\ 0 & b \end{pmatrix} \in H$, where $a \neq 0$ and $b \neq 0$. Then

$$AB^{-1} = \begin{pmatrix} 1/a & 0 \\ 0 & a \end{pmatrix} \begin{pmatrix} b & 0 \\ 0 & 1/b \end{pmatrix} = \begin{pmatrix} b/a & 0 \\ 0 & a/b \end{pmatrix} = \begin{pmatrix} 1/(a/b) & 0 \\ 0 & a/b \end{pmatrix} \in H,$$

and this completes the proof.

Exercise 4.3.32. Set

$$H = \left\{ \begin{pmatrix} \cos\alpha & -\sin\alpha \\ \sin\alpha & \cos\alpha \end{pmatrix} : \alpha \in \mathbb{R} \right\}.$$

Show that H is a group with respect to the multiplication of matrices.

Solution 4.3.32. To show that $(H, \cdot)$ is a group, we show that H is a subgroup of $GL_2(\mathbb{R})$ in lieu. First, H is indeed a subset of $GL_2(\mathbb{R})$ because the determinant of any element A of H is

$$\det A = \cos\alpha \cos\alpha - (-\sin\alpha)\sin\alpha = \cos^2\alpha + \sin^2\alpha = 1,$$

making any A invertible. Besides, H does contain the identity matrix of $GL_2(\mathbb{R})$, as this corresponds to the value $\alpha = 0 \in \mathbb{R}$.

To conclude, let $A, B \in H$, i.e. $A = \begin{pmatrix} \cos\alpha & -\sin\alpha \\ \sin\alpha & \cos\alpha \end{pmatrix}$ and $B = \begin{pmatrix} \cos\beta & -\sin\beta \\ \sin\beta & \cos\beta \end{pmatrix}$, where $\alpha, \beta \in \mathbb{R}$. By Proposition 4.1.8, we know the expression of B^{-1}. Since

$$AB^{-1} = \begin{pmatrix} \cos\alpha & -\sin\alpha \\ \sin\alpha & \cos\alpha \end{pmatrix} \begin{pmatrix} \cos\beta & +\sin\beta \\ -\sin\beta & \cos\beta \end{pmatrix},$$

making use of well-known trigonometric identities, as well as the fact that the cosine function is even and the sine function is odd, we finally obtain that

$$AB^{-1} = \begin{pmatrix} \cos(\alpha-\beta) & -\sin(\alpha-\beta) \\ \sin(\alpha-\beta) & \cos(\alpha-\beta) \end{pmatrix}.$$

In other words, AB^{-1} too is of the type of elements in H, and this ends the proof.

Exercise 4.3.33. Let $(G, *)$ be a group. We call the center of G the set defined by

$$C_G = \{x \in G : x * y = y * x, \ \forall y \in G.\}$$

(1) Prove that C_G is a subgroup of G.
(2) What can we say about C_G in the case $(G, *)$ is abelian?

Solution 4.3.33.

(1) First, observe that $e \in C_G$. Let $x, x' \in C_G$. We have to show that $x * x'^{-1} \in C_G$. Let $y \in G$. Since $x, x' \in C_G$, we get

$$x * y = y * x \text{ and } x' * y = y * x'.$$

So

$$x * x'^{-1} * y = x * y * x'^{-1} = y * x * x'^{-1},$$

i.e. $x * x'^{-1} \in C_G$.

(2) If $(G, *)$ is commutative, then

$$\forall x, y \in G : x * y = y * x.$$

In other words, $C_G = G$.

Exercise 4.3.34. Let G be a group, and let $a \in G$. Show that $H := \{a^n : n \in \mathbb{Z}\}$ is a subgroup of G (this is called a cyclic subgroup generated by x).

Solution 4.3.34. Denote the binary operation on G by "$\cdot$". First, H is not empty, as it contains a, which corresponds to $n = 1$. Let $x, y \in G$, i.e. $x = a^n$ and $y = a^m$ for some integers n and m. By Proposition 4.1.3, we may write

$$x \cdot y^{-1} = a^n \cdot (a^m)^{-1} = a^n \cdot a^{-m} = a^{n-m} \in H$$

as $n - m \in \mathbb{Z}$. Consequently, H is a subgroup of G, as suggested.

Exercise 4.3.35. Let $(G, *)$ be a group with e as its identity element, and let $H \subsetneq G$ be a subgroup of G, and let H^c be the complement of H in G. Set

$$H^c * H = \{g * h : g \in H^c, h \in H\}$$

and

$$H * H^c = \{h * g : h \in H, g \in H^c\},$$

where the compositions $h * g$ and $g * h$ are carried out in $(G, *)$. Determine whether of the following inclusions are true ("false" requires a counterexample, and "true" requires a proof):

(1) $H * H \subset H$,
(2) $H^c * H^c \subset H^c$,
(3) $H * H^c \subset H^c$,
(4) $H^c * H \subset H^c$,
(5) $H^c * H \subset H$,
(6) $H * H^c \subset H$.

Solution 4.3.35.

(1) True! Let $x \in H * H$, i.e. $x = h * k$ where $h, k \in H$. Since H is a subgroup, $x = h * k \in H$.
(2) False! If $* = +$, $G = \mathbb{R}$, and $\mathbb{Q} = H$, then $-\sqrt{2}, \sqrt{2} \in \mathbb{Q}^c$, but $\sqrt{2} + (-\sqrt{2}) = 0 \notin \mathbb{Q}^c$. In other words, $\mathbb{Q}^c + \mathbb{Q}^c \not\subset \mathbb{Q}^c$.
(3) True. For a proof, let $k = h * g \in H * H^c$, where $h \in H$ and $g \in H^c$. If $k = h * g \in H$, then it would follow that

$$g = e * g = h^{-1} * h * g \in H,$$

which is absurd. Therefore, $k \in H^c$, as required.
(4) True (reason as before).
(5) False! For instance, $1 \in \mathbb{Q}$ and $\sqrt{2} \in \mathbb{Q}^c$, but $\sqrt{2} + 1 \notin \mathbb{Q}$. So, $\mathbb{Q}^c + \mathbb{Q} \not\subset \mathbb{Q}$.
(6) False (argue as before).

Exercise 4.3.36. Let $(G, *)$ be a group and let e be its identity element. Suppose H and K are two subgroups of G.

(1) Show that $H \cap K$ remains a subgroup of G.
(2) Is the union $H \cup K$ always a subgroup of G?
(3) Show that $H \cup K$ is a subgroup if and only if $H \subset K$ or $K \subset H$.

Solution 4.3.36.

(1) First, observe that $H \cap K$ is non-empty, and more precisely $e \in H \cap K$. Now, let $h, k \in H \cap K$, and so $h, k \in H$ and

$h, k \in K$. Since H is a subgroup, $h * k^{-1} \in H$, and since K is a subgroup, $h * k^{-1} \in K$. Therefore, $h * k^{-1} \in H \cap K$.

(2) The answer is negative. Let $H = 2\mathbb{Z}$ and $K = 3\mathbb{Z}$. Then both H and K are subgroups of $(\mathbb{Z}, +)$. However, $(2\mathbb{Z} \cup 3\mathbb{Z}, +)$ is not a subgroup of $(\mathbb{Z}, +)$ as, e.g. $2 \in H \subset H \cup K$ and $3 \in K \subset H \cup K$, whereas $2 + 3 = 5 \notin H \cup K$.

(3) If either $H \subset K$ or $K \subset H$, there is nothing to prove as $K \cup H$ will be either K or H. Conversely, suppose that $H \cup K$ is a group. For the sake of contradiction, we assume that $H \not\subset K$ *and* $K \not\subset H$. Hence there is some h in H such that $h \notin K$, and there is a k in K such that $k \notin H$. So, $h \in H \cup K$ and $k \in H \cup K$. Since it is assumed that $H \cup K$ is a subgroup, it is seen that $h * k \in H \cup K$. Therefore, either $h * k \in H$ or $h * k \in K$. We will see that each case yields a contradiction. For instance, when $h * k \in H$ then $h^{-1} * h * k \in H$ or $e * k = k \in H$, which is impossible. Similarly, if $h * k \in K$, then $h = h * e = h * k * k^{-1} \in K$, another contradiction. This marks the end of the proof.

Exercise 4.3.37. ([3]) Show that a group is never the union of two proper subgroups.

Solution 4.3.37. Let G be a group, and let H and K be two subgroups of G such that $H \neq G$ and $K \neq G$. We are requested to show that $G \neq H \cup K$. We argue by contradiction, i.e. assume that $G = H \cup K$. Hence $H \not\subset K$ and $K \not\subset H$. So, choose an $h \in H$ that is not in K, and a $k \in K$ which is not in H. Obviously, $hk \in G$ and $hk \notin H$. Indeed, if $hk \in H$, then $k = h^{-1}hk \in H$, and this is untrue. Similarly, $hk \notin K$. So, $hk \notin H \cup K$, and so $G \neq H \cup K$, as wished.

Exercise 4.3.38. Let $(G, *)$ and $(H, \perp)$ be two groups. There is a straightforward way for defining a binary operation, denoted by "$\bullet$", on the direct (or cartesian) product $G \times H$: If $x, x' \in G$ and $y, y' \in H$, then put

$$(x, y) \bullet (x', y') = (x * x', y \perp y').$$

Show that $(G \times H, \bullet)$ is a group. Show that if G and H are abelian if and only if $G \times H$ is.

Solution 4.3.38. There is not much to prove. One just needs to write the definitions correctly and appropriately, then one uses the fact that G and H are groups, and everything will fit perfectly! Let interested readers fill in all gaps.

The identity element is (e_G, e_H), where e_G and e_H denote the identity elements of G and H respectively, while the inverse of each (x, y) is (x^{-1}, y^{-1}).

Regarding commutativity, let $x, x' \in G$ and $y, y' \in H$. Then, it is seen that if G and H are abelian, $x * x' = x' * x$ and $y' \perp y = y \perp y'$. So,

$$(x, y) \bullet (x', y') = (x', y') \bullet (x, y),$$

that is, "$\bullet$" is commutative. Conversely, if "$\bullet$" is commutative, then $(x, y) \bullet (x', y') = (x', y') \bullet (x, y)$ for all $x, x' \in G$ and $y, y' \in H$. That is,

$$(x * x', y \perp y') = (x' * x, y' \perp y)$$

still for all $x, x' \in G$ and $y, y' \in H$. We then derive from the previous equality the commutativity of the operations "$*$" and "$\perp$", as wished.

Exercise 4.3.39. Let $(G, *)$ be a group. Show that the set

$$H = \{(a, a) : a \in G\}$$

is a subgroup of $(G \times G, \bullet)$.

Solution 4.3.39. First, H is non-empty as it contains (e, e) where e denotes the identity element of G. Now, let $(a, a), (b, b) \in H$, where $a, b \in G$. Then

$$(a, a) \bullet (b, b)^{-1} = (a, a) \bullet (b^{-1}, b^{-1}) = (a * b^{-1}, a * b^{-1}) \in H$$

for $a * b^{-1} \in G$.

Exercise 4.3.40. Determine in each case whether the given function $f : G \to H$ is a homomorphism:

(1) G is any group, $H = G$, $f(x) = e$ for all $x \in G$, where e is the identity element of G (and H!).
(2) $G = H = (\mathbb{R}, \times)$, $f(x) = x$.
(3) $G = H = (\mathbb{R}^*, \times)$, $f(x) = |x|$.
(4) $G = H = (\mathbb{R}, +)$, $f(x) = |x|$.
(5) $G = (GL_2(\mathbb{R}), \cdot)$, $H = (\mathbb{R}^*, \times)$, $f(A) = \det A$.
(6) $G = H = (\mathbb{Z}, +)$, $f(n) = n + 1$.
(7) $G = ((0, \infty), \times)$, $H = (\mathbb{R}, +)$, $f(x) = \ln x$.

In each case f is a homomorphism, check whether it is an isomorphism.

Solution 4.3.40.

(1) f is indeed a homomorphism (called the trivial homomorphism). The simple details are left to readers...
(2) f is not a group homomorphism simply because $(\mathbb{R}, \times)$ is not a group!

(3) f is a homomorphism as for all $x, y \in \mathbb{R}^*$,

$$f(xy) = |xy| = |x||y| = f(x)f(y).$$

(4) If f were a homomorphism, we would have

$$\forall x, y \in \mathbb{R} : f(x+y) = |x+y| = |x| + |y| = f(x) + f(y).$$

This is not always true (witness $x = 1$ and $y = -3$).

(5) f is a homomorphism as we have already observed that

$$\det(AB) = \det A \det B$$

for all $A, B \in M_2(\mathbb{R})$, in particular in $GL_2(\mathbb{R})$.

(6) f is not a homomorphism because, e.g.

$$f(0) = 1 \neq 0.$$

(7) f is a homomorphism for all $x, y > 0$

$$f(xy) = \ln(xy) = \ln x + \ln y = f(x) + f(y).$$

We have seen that the functions in (1), (3), (5) and (7) are homomorphisms:

- The function in (1) is not one-to-one, so f is not an isomorphism.
- The function in (3) is not one-to-one (and not onto either), so f is not an isomorphism.
- The function in (5) is not one-to-one: Indeed, if $A = \begin{pmatrix} 1 & 0 \\ 0 & 2 \end{pmatrix}$ and $B = \begin{pmatrix} 4 & 0 \\ 0 & 1/2 \end{pmatrix}$, then $A \neq B$ and yet $\det A = \det B$ $(=2)$. So, "det" is not an isomorphism.
- The function in (7) is known to be bijective, and so it is an isomorphism.

Exercise 4.3.41. Interpret the following well-known identities as morphisms of appropriate groups:

(1) $|zz'| = |z||z'|$ (where z and z' are complex numbers).
(2) $e^{x+y} = e^x e^y$.
(3) $(xy)^n = x^n y^n$, where $n \in \mathbb{Z}$.

Solution 4.3.41.

(1) Let $f : (\mathbb{C}^*, \times) \to (\mathbb{R}^*, \times)$ be defined by $f(z) = |z|$. Then f is a homomorphism of groups which satisfies:

$$f(zz') = |zz'| = |z||z'| = f(z)f(z').$$

Remark. We could have also taken f from $(\mathbb{C}^*, \times)$ into $(\mathbb{C}^*, \times)$. However, and despite the fact that the above identity is satisfied for $z = 0$, we still need to exclude this value for we need the structure of a group.

(2) Let $f : (\mathbb{R}, +) \to ((0, \infty), \times)$ be defined by $f(x) = e^x$. Then the given identity may be reformulated as the homomorphism f (indeed, $f(x + y) = f(x)f(y)$).

(3) We could use explicit groups such as $(\mathbb{R}^*, \times)$. But, more generally, let $(G, *)$ be an abelian group, and let $f : (G, *) \to (G, *)$ be defined $f(x) = x^n$, where n is a fixed integer. Then, by Exercise 4.3.27:

$$f(x * y) = (x * y)^n = x^n * y^n = f(x) * f(y)$$

and so the desired identity holds when $*$ is the usual multiplication or any binary operation noted as such.

Exercise 4.3.42. On $\mathbb{R}$, define the binary operation "$*$" by:

$$\forall(x, y) \in \mathbb{R}^2 : \quad x * y = x + y - xy.$$

(1) Show that $(\mathbb{R} - \{1\}, *)$ is a commutative group.
(2) Find an isomorphism between $(\mathbb{R} - \{1\}, *)$ and $(\mathbb{R}^*, \cdot)$.
(3) Compute

$$\underbrace{x * x * \cdots * x}_{n \text{ times}}$$

in terms of n and x.

Solution 4.3.42.

(1) The same method of Exercise 4.3.17 may be applied here to answer this question, and this is left to readers.
(2) Let $f : (\mathbb{R} - \{1\}, *) \to (\mathbb{R}^*, \cdot)$ be defined by $f(x) = 1 - x$. Then f is a group homomorphism because

$$\begin{aligned}
\forall x, y \in \mathbb{R} - \{1\} : \quad f(x * y) &= 1 - x * y \\
&= 1 - (x + y - xy) \\
&= 1 - x - y + xy \\
&= (1 - x)(1 - y) \\
&= f(x) \cdot f(y).
\end{aligned}$$

Since

$$\forall y \in \mathbb{R}^*, \ \exists x! = (1 - y) \in \mathbb{R} - \{1\} : \ f(x) = f(1 - y) = 1 - (1 - y) = y,$$

it is seen that f is also bijective, so that f becomes a group isomorphism between $(\mathbb{R} - \{1\}, *)$ and $(\mathbb{R}^*, \cdot)$.

(3) The idea is to start by calculating $x * x$ and $x * x * x$ say, then we try to guess a general formula which should work for all n, then we show that this formula is indeed true by induction. We have

$$x * x = x + x - x^2 = 2x - x^2$$

and

$$x * x * x = x * (x * x) = x * (2x - x^2) = x + 2x - x^2 - x(2x - x^2),$$

i.e.

$$x * x * x = 3x - 3x^2 + x^3.$$

Before carrying on, we remark that we have *almost* $(1 - x)^2$ in the first case, and *almost* $(1 - x)^3$ in the second case. More precisely, we have

$$x * x = 1 - (1 - x)^2$$

and

$$x * x * x = 1 - (1 - x)^3.$$

We therefore conjecture that

$$\underbrace{x * x * \cdots * x}_{n \text{ times}} = 1 - (1 - x)^n.$$

Let's show this by induction. The statement is true for $n = 1$ because

$$x = 1 - (1 - x)^1 = 1 - 1 + x = x.$$

Suppose the statement is true for n, and we show it is true for $n + 1$. We have

$$\underbrace{x * x * \cdots * x}_{n+1 \text{ times}} = x * \left(\underbrace{x * x * \cdots * x}_{n \text{ times}} \right)$$

$$= x + \underbrace{x * x * \cdots * x}_{n \text{ times}} - x \left(\underbrace{x * x * \cdots * x}_{n \text{ times}} \right)$$

$$= x + 1 - (1 - x)^n - x[1 - (1 - x)^n]$$

$$= x + 1 - (1 - x)^n - x + x(1 - x)^n$$

$$= 1 + (1 - x)^n(-1 + x)$$

$$= 1 - (1 - x)^n(1 - x)$$

$$= 1 - (1 - x)^{n+1}.$$

Remark. In principle, the method used in the last answer should always work for this type of questions. In this exercise, there is a more elegant way for obtaining the formula. We know that $f : (\mathbb{R}-\{1\}, *) \to (\mathbb{R}^*, \cdot)$ defined by $f(x) = 1 - x$ is a group isomorphism, and its inverse is given by $f^{-1}(x) = 1 - x$.

Let $n \in \mathbb{N}$. We have:

$$f(\underbrace{x * x * \cdots * x}_{n \text{ times}}) = \underbrace{f(x) \cdot f(x) \cdots f(x)}_{n \text{ times}}$$

$$= \underbrace{(1 - x)(1 - x) \cdots (1 - x)}_{n \text{ times}}$$

$$= (1 - x)^n.$$

Hence,

$$\underbrace{f^{-1}(f(x * x * \cdots * x))}_{=x*x*\cdots*x} = f^{-1}((1 - x)^n) = 1 - (1 - x)^n,$$

i.e.

$$x * x * \cdots * x = 1 - (1 - x)^n.$$

Exercise 4.3.43. We define an operation "$*$" on $\mathbb{R}$ by

$$x * y = \sqrt[3]{x^3 + y^3},$$

for all $x, y \in \mathbb{R}$.
  (1) Show that $(\mathbb{R}, *)$ is a commutative group.
  (2) Show that $(\mathbb{R}, *)$ is isomorphic to $(\mathbb{R}, +)$.

Solution 4.3.43.

  (1) (a) First, "$*$" is a binary operation because we are allowed to take the cube root of any real number, and so the closure axiom is satisfied.
      (b) Let $x, y \in \mathbb{R}$. Since "$+$" is commutative over $\mathbb{R}$, for all $x, y \in \mathbb{R}$ we have $x^3 + y^3 = y^3 + x^3$. Hence,

$$\sqrt[3]{x^3 + y^3} = \sqrt[3]{y^3 + x^3} \iff x * y = y * x,$$

i.e. we have shown that "$*$" is commutative.

(c) Let $x, y, z \in \mathbb{R}$. We have:

$$(x * y) * z = (\sqrt[3]{x^3 + y^3}) * z$$
$$= \sqrt[3]{(\sqrt[3]{x^3 + y^3})^3 + z^3}$$
$$= \sqrt[3]{(x^3 + y^3) + z^3}$$
$$= \sqrt[3]{x^3 + (y^3 + z^3)} \text{ (since ``+'' is associative on } \mathbb{R})$$
$$= \sqrt[3]{x^3 + (\sqrt[3]{y^3 + z^3})^3}$$
$$= x * (\sqrt[3]{y^3 + z^3})$$
$$= x * (y * z).$$

Whence, "$*$" is associative.

(d) Since "$*$" is commutative, to find the identity element, just consider $x * e = x$. We get

$$x * e = x \iff \sqrt[3]{x^3 + e^3} = x \iff x^3 + e^3 = x^3,$$

so $e^3 = 0$. Hence $e = 0$ is the identity element.

(e) Let $x \in \mathbb{R}$. Since "$*$" is commutative, it suffices to find an $x' \in \mathbb{R}$ such that $x * x' = e = 0$. We have

$$x * x' = 0 \iff \sqrt[3]{x^3 + x'^3} = 0 \iff x^3 + x'^3 = 0 \iff x^3 = -x'^3$$

and so $x' = -x \in \mathbb{R}$. Therefore, any real number x admits an inverse $-x$. Thus, $(\mathbb{R}, *)$ is an abelian group.

(2) Let $f : (\mathbb{R}, *) \to (\mathbb{R}, +)$ be defined by $f(x) = x^3$. First, f is a bijective function. In addition, we have for all x and y:

$$f(x * y) = (x * y)^3 = (\sqrt[3]{x^3 + y^3})^3 = x^3 + y^3 = f(x) + f(y),$$

i.e. f is a group homomorphism, thereby we conclude that f is a group isomorphism from $(\mathbb{R}, *)$ onto $(\mathbb{R}, +)$, as suggested.

Exercise 4.3.44. Let $\alpha \in \mathbb{R}$. Let $f_\alpha : \mathbb{R}^2 \to \mathbb{R}^2$ be a map defined by

$$f_\alpha(x, y) = (x, y + \alpha x).$$

(1) Prove that f_α is bijective and give its inverse f_α^{-1}.
(2) Let $G = \{f_\alpha : \alpha \in \mathbb{R}\}$. Show that $(G, \circ)$ is an abelian group, where $\circ$ denotes the composition of functions.
   Hint: Show that G is a subgroup of another group!
(3) Let F be a map defined from $(\mathbb{R}, +)$ to $(G, \circ)$ by $F(\alpha) = f_\alpha$ for all $\alpha \in \mathbb{R}$. Prove that F is an isomorphism.

Solution 4.3.44.

(1) Let $\alpha \in \mathbb{R}$. First, we show that f_α is injective. Let (x, y) and (x', y') be both in $\mathbb{R}^2$ and suppose that

$$f_\alpha(x, y) = f_\alpha(x', y'), \text{ i.e. } (x, y + \alpha x) = (x', y' + \alpha x').$$

Then,

$$x = x' \text{ and } y + \alpha x = y' + \alpha x'$$

and so $y = y'$ too. Hence, $(x, y) = (x', y')$, i.e. f is injective.

Now, we show that f_α is surjective. Let $(t, z) \in \mathbb{R}^2$ be such that $f_\alpha(x, y) = (t, z)$, i.e. $(x, y + \alpha x) = (t, z)$. Thus,

$$\begin{cases} x = t, \\ y + \alpha x = z, \end{cases}$$

which gives

$$x = t \text{ and } y = z - \alpha x = z - \alpha t.$$

Whence, f_α is surjective and consequently, f_α is bijective. By the previous calculation we even have the expression of f_α^{-1}, namely:

$$f_\alpha^{-1}(x, y) = (x, y - \alpha x).$$

The observation

$$f_\alpha^{-1} = f_{-\alpha}$$

will be very useful for what remains of the exercise.

(2) Since G is a subset of the set of bijective functions of $\mathbb{R}^2$ onto $\mathbb{R}^2$ (denoted by $\mathcal{B}$), and since this latter is a group with respect to the composition of functions (see the remark in Answer 4.3.13, with $E = \mathbb{R}^2$), we show that G is a subgroup of $\mathcal{B}$. The identity element of $\mathcal{B}$, i.e. $\mathrm{id}_{\mathbb{R}^2}$ belongs to G (take $\alpha = 0$).

Let $\alpha, \beta \in \mathbb{R}$. We show that $f_\alpha \circ f_\beta^{-1} \in G$. We have for all $(x, y) \in \mathbb{R}^2$

$$\begin{aligned} f_\alpha \circ f_\beta^{-1}(x, y) &= f_\alpha \circ f_{-\beta}(x, y) \\ &= f_\alpha(f_{-\beta}(x, y)) \\ &= f_\alpha(x, y - \beta x) \\ &= (x, y - \beta x + \alpha x) \\ &= (x, y + (-\beta + \alpha)x). \end{aligned}$$

The latter is clearly of the same form of the elements of G (notice that $f_\alpha \circ f_\beta^{-1} = f_{\alpha - \beta}$), and so $f_\alpha \circ f_\beta^{-1} \in G$. Thus, $(G, \circ)$ is indeed a group.

(3) The map F is obviously onto. Let's show that it is one-to-one. Let $\alpha, \beta \in \mathbb{R}$ be such that $F(\alpha) = F(\beta)$, i.e.

$$f_\alpha(x, y) = f_\beta(x, y), \ \forall(x, y) \in \mathbb{R}^2,$$

i.e.

$$(x, y + \alpha x) = (x, y + \beta x), \ \forall(x, y) \in \mathbb{R}^2.$$

Hence

$$\alpha x = \beta x, \ \forall x \in \mathbb{R},$$

i.e. $\alpha = \beta$. This proves that F is injective. Therefore, F is bijective. There only remains to show that F is a homomorphism between $(\mathbb{R}, +)$ and $(G, \circ)$, i.e.

$$\forall \alpha, \beta \in \mathbb{R} : \ F(\alpha + \beta) = F(\alpha) \circ F(\beta).$$

But this true as we have already seen that $f_\alpha \circ f_\beta = f_{\alpha+\beta}$.

Exercise 4.3.45. Let $(G, *)$ be a group and let $a \in G$ be given. Show that the map $f : (G, *) \to (G, *)$ defined by $f(x) = a * x * a^{-1}$ is an automorphism.

Solution 4.3.45. First, we show that f is a morphism. Let $x, y \in G$. Then we have

$$f(x * y) = a * x * y * a^{-1} = (a * x * a^{-1}) * (a * y * a^{-1}) = f(x) * f(y).$$

Now, we prove that f is injective. We may show that $\ker f = \{e\}$. We have

$$\ker f = \{x \in G : \ a * x * a^{-1} = e\},$$

but

$$a * x * a^{-1} = e \iff a * x = e * a = a \iff x = a^{-1} * a = e.$$

Hence $\ker f \subset \{e\}$. Since $e \in \ker f$ for all morphisms, we immediately deduce that

$$\ker f = \{e\},$$

i.e. f is injective. Next, we show that f is surjective. Let $y \in G$ be such that $f(x) = y$, i.e. $a * x * a^{-1} = y$. For $x = a^{-1} * y * a \in G$, we have

$$f(x) = f(a^{-1} * y * a) = a * a^{-1} * y * a * a^{-1} = y.$$

Consequently, f is bijective, and so it is an isomorphism from G to G. Thus, f is indeed an automorphism.

Exercise 4.3.46. Let $(G, *)$ be a group, and let e be its identity element. Then define a function $f : G \to G$ by $f(x) = x^{-1}$, where x^{-1} designates the inverse of x.

(1) Show that f is bijective.
(2) Under what conditions f becomes an automorphism?

Solution 4.3.46.

(1) Let $y \in G$ be such that $f(x) = y$, i.e. $x^{-1} = y$. So, $x = y^{-1} \in G$. Since inverses are unique, f is bijective.
(2) Since f is already bijective and $f : G \to G$, f is an automorphism as soon as f is a homomorphism. Let us see when f is a homomorphism: Let $x, y \in G$. Then f must satisfy

$$f(x * y) = f(x) * f(y) \text{ or } (x * y)^{-1} = x^{-1} * y^{-1}.$$

But, we already know that $(x * y)^{-1} = y^{-1} * x^{-1}$. It thus becomes apparent that f is a homomorphism if and only if

$$x^{-1} * y^{-1} = y^{-1} * x^{-1}$$

for all $x, y \in G$. A right "multiplication" of the previous equation by y, and a left "multiplication" still by y yield $y * x^{-1} = x^{-1} * y$. Proceeding as before with x implies $x * y = y * x$. Thus, f is an automorphism if and only if G is an abelian group.

4.4. Supplementary Exercises

Exercise 4.4.1. Determine whether the binary operation $*$ defined over E satisfies the closure, commutativity and associativity lass in each of the following cases:

(1) $x * y = xy + 1$, $E = \mathbb{Q}$.
(2) $x * y = \frac{xy}{2}$, $E = \mathbb{Q}$.
(3) $x * y = x^y$, $E = \mathbb{N}$.
(4) $x * y = 2^{xy}$, $E = \mathbb{N}$.

Exercise 4.4.2. Let X be a non-empty set. Let $G = \mathcal{P}(X)$ be the powerset of X. Set:

$$\forall A, B \in G : A * B = A \cap B \text{ (the intersection of } A \text{ and } B).$$

Is $(G, *)$ a group?

Exercise 4.4.3. Can you find some binary operation, noted "$*$", such that $(\mathbb{N} \cup \{0\}, *)$ becomes a group?

Exercise 4.4.4. Let $G = \{f, g, h, i, j, k\}$ where f, g, h, i, j and k be functions defined from $\mathbb{R}^* - \{1\}$ into $\mathbb{R}^* - \{1\}$ by

$$f(x) = x, \; g(x) = 1 - x, \; h(x) = \frac{1}{1-x}, \; i(x) = \frac{1}{x},$$

$$j(x) = \frac{x}{x-1} \text{ and } k(x) = \frac{x-1}{x}.$$

respectively. Let "∘" be the usual composition of functions.

(1) Prove that $(G, \circ)$ is a group.
(2) Is it commutative?

Exercise 4.4.5. On $E = (0, +\infty)$, define

$$x * y = x^{\ln y}, \; \forall x, y \in E.$$

Is $(E, *)$ an abelian group?

Exercise 4.4.6. We define an operation "$*$" over $\mathbb{R}$ by

$$x * y = x\sqrt{1 + y^2} + y\sqrt{1 + x^2},$$

for all real x and y.

(1) Prove that $(\mathbb{R}, *)$ is a commutative group.
(2) Prove that $(\mathbb{R}, *)$ is isomorphic to $(\mathbb{R}, +)$.

Exercise 4.4.7. We define "$*$" over the real numbers by

$$x * y = xy + (x^2 - 1)(y^2 - 1)$$

for all $x, y \in \mathbb{R}$. Is $(\mathbb{R}, *)$ a group?

Exercise 4.4.8. Let G and H be groups, and let K be a subgroup of $G \times H$. Is K necessarily of the form $X \times Y$ for some subgroup X of G and some subgroup Y of H?

Exercise 4.4.9. Give an example of a finite group (i.e. it contains only a finite number of elements) which is not abelian.

Exercise 4.4.10. Determine whether the reverse inclusions hold in Exercise 4.3.35.

Exercise 4.4.11. Let $f : E \to F$ be a function. Show that the following statements are true:

(1) f is injective if and only if it is left invertible.
(2) f is surjective if and only if it is right invertible.

Exercise 4.4.12. Set

$$G = \{f : \mathbb{C} \to \mathbb{C} : z \mapsto f(z) = z + n : n \in \mathbb{Z}\},$$

with $z \in \mathbb{C}$. Show that $(G, \circ)$ is a group.

Exercise 4.4.13. (Cf. Exercise 4.3.29) Let $(G, *)$ be a group such that card $G = 4$. Find its Cayley table.

Exercise 4.4.14. Show that the map $f : (\mathbb{R}, +) \to (\mathbb{C}^*, \times)$ defined by $f(x) = e^{(1+i)x}$ is a homomorphism.

CHAPTER 5

Rings and Fields

5.1. Basics

DEFINITION 5.1.1. A ring is a non-empty set R equipped with two binary operations denoted by "$+$" and "$\cdot$" (not necessarily the usual addition and multiplication) such that:

(1) $(R, +)$ is an abelian group.
(2) R is closed under "$\cdot$", i.e. for all $x, y \in R$: $x \cdot y \in R$.
(3) The binary operation "$\cdot$" is associative.
(4) (Distributive law) For all $x, y, z \in R$:

$$(x + y) \cdot z = x \cdot z + y \cdot z \text{ and } z \cdot (x + y) = z \cdot x + z \cdot y.$$

The usual notation for a ring is $(R, +, \times)$.

If "$\cdot$" is commutative, then the ring $(R, +, \cdot)$ is said to be commutative.

If there is $1_R \in R$ such that $1_R \neq 0_R$ (where 0_R is the identity element with respect to "$+$") and

$$x \cdot 1_R = 1_R \cdot x = x,$$

then $(R, +, \cdot)$ is called a ring with identity (which is 1_R).

Remark. In some references, a ring does not necessarily have to have an identity element with respect to the second binary operation, whereas in others, a ring must possess an identity element (with respect to the second operation). So, when readers use another reference, they should check which convention is being used.

Remark. As it is customary, we may just drop the dot in $a \cdot b$, and merely write it as ab by remembering that this is still the second binary operation of the ring (which could represent any operation). We will do so whenever there is no risk of confusion.

EXAMPLE 5.1.1. By the basic axioms for addition and multiplication, $(\mathbb{Z}, +, \times)$, $(\mathbb{Q}, +, \times)$, $(\mathbb{R}, +, \times)$ and $(\mathbb{C}, +, \times)$ are commutative rings with identity.

EXAMPLE 5.1.2. Let $M_2(\mathbb{R})$ be the set of 2×2 real matrices. Let "$+$" denote the usual addition of matrices, and let "$\cdot$" denote the multiplication of matrices. Then $(M_2(\mathbb{R}), +, \cdot)$ is a ring with identity. See Exercise 5.3.7 for a proof.

EXAMPLE 5.1.3. Let $\mathbb{R}[X]$ be the set of polynomials with real coefficients. Then $(\mathbb{R}[X], +, \cdot)$ is a ring. A proof may be consulted in Exercise 5.3.32.

EXAMPLE 5.1.4. (See Exercise 5.3.34) Let $n \in \mathbb{N}$. Define on $\mathbb{Z}_n$

$$\overline{x} \oplus \overline{y} = \overline{x + y},$$

and

$$\overline{x} \otimes \overline{y} = \overline{x \cdot y},$$

where $x, y \in \mathbb{Z}$. Then $(\mathbb{Z}_n, \oplus, \otimes)$ is a ring.

If we want to show that a non-empty set R, equipped with two binary operations, is a ring, and in order to avoid checking all the properties in the definition of a ring, we usually try to see whether R may be regarded as a subset of a set which is already known to be a ring (with respect to the same binary operations). This gives rise to the concept of a subring:

DEFINITION 5.1.2. (Subring test) Let $(R, +, \cdot)$ be a ring and let S be a non-empty subset of R. We say that S is a subring of R if:

$$\forall x, y \in S : \ x + (-y) = x - y \in S \text{ and } x \cdot y \in S.$$

EXAMPLE 5.1.5. $2\mathbb{Z}$ is a subring of $(\mathbb{Z}, +, \cdot)$. See Exercise 5.3.8 for a proof.

We recall two more important notions:

DEFINITION 5.1.3. Let $(R, +, \cdot)$ be a ring with identity. We say that R is an integral domain if

$$\forall a, b \in R : a \cdot b = 0_R \implies a = 0_R \text{ or } b = 0_R,$$

where 0_R is the identity element with respect to "$+$".

Remark. Observe that in the above definition, like many references (and unlike many others as well), we are not requiring the ring to be commutative.

Next, we define ring homomorphisms.

DEFINITION 5.1.4. Let $(R, +, \cdot)$ and $(S, *, \bullet)$ be two rings, and let $f : R \to S$ be a certain mapping. We say that f is a ring homomorphism if for all $x, y \in R$:

$$f(x + y) = f(x) * f(y) \text{ and } f(x \cdot y) = f(x) \bullet f(y).$$

In case f is also bijective, the homomorphism f is called an isomorphism.

Remark. References in which a ring is required to have an identity element, usually add the condition $f(1_R) = 1_S$ in the definition of a ring homomorphism.

Finally, we recall the definition of a field.

DEFINITION 5.1.5. A field $(F, +, \cdot)$ is a ring with identity, in which every non-zero element has an inverse.

Equivalently, $(F, +, \cdot)$ is a field if and only if the following three conditions hold:

   (1) $(F, +)$ is an abelian group,
   (2) $(F \setminus \{0\}, \cdot)$ is a group,
   (3) and the distributive laws are satisfied.

When the second binary operation is commutative, then we may speak of a commutative field.

EXAMPLE 5.1.6. $(\mathbb{Q}, +, \times)$, $(\mathbb{R}, +, \times)$ and $(\mathbb{C}, +, \times)$ are commutative fields.

EXAMPLE 5.1.7. $(\mathbb{Z}, +, \times)$ is not a field. Indeed, 2 does not have a multiplicative inverse as $1/2 \notin \mathbb{Z}$.

EXAMPLE 5.1.8. (See Exercise 5.3.34) $(\mathbb{Z}_n, \oplus, \otimes)$ is a field if and only if n is a prime number.

Remark. Readers are reminded once more that they should be careful with other references when it comes to fields. Indeed, some define a field to be a commutative ring with identity.

DEFINITION 5.1.6. Let F and F' be two fields. Say that $f : F \to F'$ be a field homomorphism (resp. isomorphism) when it is a ring homomorphism (resp. isomorphism).

5.2. True or False

Questions. Determine, giving reasons, whether the following statements are true or false.

(1) If $(R, +, \cdot)$ is a ring with identity element 1_R, where 0_R denotes the identity element with respect to "$+$", and if $1_R = 0_R$, then R is reduced to the singleton $\{0_R\}$.

(2) Can we have algebraic structures with three binary operations?

(3) Let $(G, +)$ be some abelian group with the identity element denoted by 0. Then we may always define a certain multiplication, noted "$\cdot$", such that $(G, +, \cdot)$ becomes a ring.

(4) In $\mathbb{R}$, the addition is distributive over multiplication.

(5) $(M_2(\mathbb{R}), \cdot, +)$ is a ring.

(6) If $(R, +, \cdot)$ is a ring and $a, b \in A$. Complete the following expressions:

$$(a + b)(a - b) = \cdots$$

and

$$(a + b)^3 = \cdots$$

(7) (Freshman's dream) Assume that R is a commutative ring such that $x + x = 0$ for all $x \in R$. Then

$$(x + y)^2 = x^2 + y^2$$

for all $x, y \in R$.

(8) In the ring $(\mathbb{Z}, +, \cdot)$, if $a, b \in \mathbb{Z}$ are such that $ab = 1$, then $a = \pm 1$ and $b = \pm 1$ respectively.

(9) In a ring $(R, +, \cdot)$, if $a \in R$ and $a^2 = 1$, then either $a = 1$ or $a = -1$.

(10) In an integral domain $(R, +, \cdot)$, if $a \in R$ and $a^2 = 1$, then either $a = 1$ or $a = -1$.

(11) In a ring $(R, +, \cdot)$, if $a \in R$ and $a^2 = a$, then either $a = 1$ or $a = 0$.

(12) Let a and b be elements of some integral domain $(R, +, \cdot)$. If $a + b - ab = 1$, then either $a = 1$ or $b = 1$.

(13) Let $(R, +, \cdot)$ be a ring. If $a \in R$ is such that $a^2 = 0_R$, then $a = 0_R$.

(14) Let $(R, +, \cdot)$ be a ring and let $a, b \in R$. Then

$$ab = 0_R \implies ba = 0_R.$$

(15) Let $(R, +, \cdot)$ be a ring and let $a, b \in R$. Then

$$ab = ba \iff a^2 b = ba^2.$$

(16) If $(R, +, \cdot)$ is a ring and $x \in R$ is such that $2x = 0$, then $x = 0$.

(17) (Cf. Exercises 5.3.35 & 5.3.23) Let F be a finite field. Then $\sum_{x \in F} x = 0$.

(18) Let $(R, +, \cdot)$ be a ring and let $a, b \in R$ be two invertible elements with respect to "$\cdot$". Is $a + b$ necessarily invertible with respect to "$\cdot$"?

(19) If $(R, +, \cdot)$ is a ring with identity, so is any subring of it.

(20) If $f : (R, +, \cdot) \to (R', +, \cdot)$ is a ring homomorphism, then $f(0_R) = 0_{R'}$, where 0_R and $0_{R'}$ are the identity elements of $(R, +)$ and $(R', +)$ respectively.

(21) In any ring homomorphism $f : R \to S$, then $f(-x) = -f(x)$ for any $x \in R$.

(22) If $f : R \to S$ is a ring homomorphism, then $f(1_R) = 1_S$ where 1_R and 1_S are the identity elements of R and S respectively.

(23) We have not imposed the condition $f(1) = 1$ in the definition of a ring homomorphism $f : R \to S$. However, if f is a *field isomorphism*, then $f(1) = 1$.

Answers.

(1) True. As a simple proof, notice that when $0_R = 1_R$, and if $x \in R$, then

$$0_R \cdot x = 1_R \cdot x \implies 0_R = x,$$

i.e. $R \subset \{0_R\}$. Since $0_R \in R$ (by definition of a ring), we infer that $R = \{0_R\}$, as wished.

Remark. It is noteworthy that the identity $0_R \cdot x = 0$ does require a proof (see, e.g. Exercise 5.3.1).

(2) Yes, we can. For example, I. Adler (he is not the only one) introduced in [1] the so-called composition rings. He observed that rings of polynomials may be equipped with the operations: "+", "$\cdot$", and the usual composition "$\circ$" which always obeys $(p + q) \circ r = p \circ r + q \circ r$, $(pq) \circ r = (p \circ r)(q \circ r)$ and $(p \circ q) \circ r = p \circ (q \circ r)$, where p, q and r are polynomials. This was his point of departure for defining more general algebraic structures with three binary operations.

(3) True. Define the trivial binary operation $x \cdot y = 0$ for all $x, y \in G$. This rule satisfies the laws of the second operation in the definition of a ring. For example, for all $x, y \in G$: $x \cdot y = 0 \in G$, showing the closure law. To see why "$\cdot$" is associative, let $x, y, z \in G$. Then

$$(x \cdot y) \cdot z = 0 \cdot z = 0 = x \cdot (y \cdot z).$$

The other laws are left to readers for verification.

(4) False! We are asked whether we have

$$x + (yz) = (x + y)(x + z), \ \forall x, y, z \in \mathbb{R}.$$

But, this is untrue! For instance, if $x = 1$, $y = 2$ and $z = 0$, then

$$1 + 2 \times 0 = 1 \neq (1 + 2)(1 + 0) = 3.$$

(5) False! One reason (among others) which prevents $(M_2(\mathbb{R}), \cdot, +)$ from being a ring is that $(M_2(\mathbb{R}), \cdot)$ is not a commutative group, as already observed before.

(6) We have:

$$(a + b)(a - b) = a^2 + ba - ab - b^2.$$

However, if the two particular elements a and b commute, that is, $ab = ba$ (and this a lot less restrictive than requiring the entire ring to be commutative), then

$$(a + b)(a - b) = a^2 - b^2.$$

For the other identity, we have

$$(a + b)^3 = a^3 + a^2b + aba + ba^2 + ab^2 + bab + b^2a + b^3.$$

But if $ab = ba$, then

$$(a + b)^3 = a^3 + 3a^2b + 3ab^2 + b^3.$$

This applies to the binomial theorem which reads: If $(R, +, \cdot)$ is a ring and $a, b \in R$ *commute*, i.e. $ab = ba$, then

$$(a + b)^n = \sum_{k=0}^{n} \binom{n}{k} a^{n-k} b^k$$

for all $n \in \mathbb{N}$, where $\binom{n}{k} = \frac{n!}{k!(n-k)!}$. Readers may prove it using a proof by induction. Let us give a counterexample which shows the failure of the previous identity in the case of the absence of the condition $ab = ba$. An efficient place to find a counterexample is the ring $(M_2(\mathbb{R}), +, \cdot)$, and take $n = 2$. Let $A = \begin{pmatrix} 0 & 1 \\ 0 & 0 \end{pmatrix}$ and $B = \begin{pmatrix} 0 & 0 \\ 1 & 0 \end{pmatrix}$. Readers may readily check that

$$AB = \begin{pmatrix} 1 & 0 \\ 0 & 0 \end{pmatrix} \neq \begin{pmatrix} 0 & 0 \\ 0 & 1 \end{pmatrix} = BA,$$

that $A + B = \begin{pmatrix} 0 & 1 \\ 1 & 0 \end{pmatrix}$, and that $A^2 = B^2 = 0_{M_2(\mathbb{R})}$. Now,

$$(A + B)^2 = \begin{pmatrix} 0 & 1 \\ 1 & 0 \end{pmatrix}^2 = \begin{pmatrix} 1 & 0 \\ 0 & 1 \end{pmatrix}$$

which differs from

$$A^2 + 2AB + B^2 = \begin{pmatrix} 2 & 0 \\ 0 & 0 \end{pmatrix}.$$

(7) True (cf. Exercise 5.4.14). Before giving a proof, recall that a ring with identity is said to have characteristic $p \in \mathbb{N}$ when p is the smallest number such that

$$\underbrace{1 + 1 + \cdots + 1}_{p \text{ times}} = 0.$$

In symbols, $\mathrm{char}(R) = p$. In case a ring does not have a characteristic, then set $\mathrm{char}(R) = 0$ (e.g. $\mathrm{char}(\mathbb{R}) = \mathrm{char}(\mathbb{Q}) = 0$).

So, our question is about rings of characteristic 2. Let us give a proof of the statement: First, we know that

$$(x + y)^2 = x^2 + xy + yx + y^2$$

for all $x, y \in R$. Since R is commutative, $xy = yx$, and so $xy + yx = xy + xy = 0$ since $\mathrm{char}(R) = 2$. Thus,

$$(x + y)^2 = x^2 + y^2,$$

as required.

(8) True. Let's give a proof. Since $ab = 1$, clearly $a \neq 0$ and $b \neq 0$. So, e.g. $a = 1/b$. Since $a \in \mathbb{Z}$, $1/b \in \mathbb{Z}$ which is only possible when $b = \pm 1$, whereby $a = \pm 1$.

(9) False! Consider the ring $(M_2(\mathbb{R}), +, \cdot)$. Its identity element is $I_2 = \begin{pmatrix} 1 & 0 \\ 0 & 1 \end{pmatrix}$. If $A^2 = I_2$, then $A = -I_2$ and $A = I_2$ do satisfy the previous equation as

$$(-I_2)^2 = (-I_2)(-I_2) = \begin{pmatrix} -1 & 0 \\ 0 & -1 \end{pmatrix}\begin{pmatrix} -1 & 0 \\ 0 & -1 \end{pmatrix} = \begin{pmatrix} 1 & 0 \\ 0 & 1 \end{pmatrix}.$$

Similarly, it is seen that $I_2^2 = I_2$. However, these are not the only possible choices. For example, there are two other obvious instances, namely: $\begin{pmatrix} 1 & 0 \\ 0 & -1 \end{pmatrix}$ and $\begin{pmatrix} -1 & 0 \\ 0 & 1 \end{pmatrix}$ whose squares also gives I_2. In fact, the equation $A^2 = I_2$ has an infinitude of solutions given by

$$A_x = \begin{pmatrix} x & 1 \\ 1 - x^2 & -x \end{pmatrix},$$

where $x \in \mathbb{R}$. Indeed, for all $x \in \mathbb{R}$, we have:

$$A_x^2 = \begin{pmatrix} x & 1 \\ 1 - x^2 & -x \end{pmatrix} \begin{pmatrix} x & 1 \\ 1 - x^2 & -x \end{pmatrix} = \begin{pmatrix} x^2 + 1 - x^2 & x - x \\ x - x^3 - x + x^3 & 1 - x^2 + x^2 \end{pmatrix}$$

and so $A_x^2 = I_2$ whichever x.

Remark. The hidden notion in the preceding question is the so-called square root. See Exercise 5.3.42 for more about this concept in rings.

As for matrices (or operators), and if some students ever take a course on advanced matrix theory (or operator theory), they will be able to consult, e.g. [**23**] or [**24**], for more about this topic.

(10) True. To see why, let a be in R and such that $a^2 = 1$. Then

$$0 = a^2 - 1 = (a - 1)(a + 1).$$

Since $a - 1, a + 1 \in R$ and R is an integral domain, it follows that $a - 1 = 0$ or $a + 1 = 0$, i.e. $a = 1$ or $a = -1$.

(11) False. Consider the ring $(M_2(\mathbb{R}), +, \cdot)$. If $A = \begin{pmatrix} 1 & 0 \\ 0 & 0 \end{pmatrix}$, then

$$A^2 = \begin{pmatrix} 1 & 0 \\ 0 & 0 \end{pmatrix} \begin{pmatrix} 1 & 0 \\ 0 & 0 \end{pmatrix} = \begin{pmatrix} 1 & 0 \\ 0 & 0 \end{pmatrix} = A,$$

and yet

$$A \neq \begin{pmatrix} 0 & 0 \\ 0 & 0 \end{pmatrix} = 0_{M_2(\mathbb{R})} \text{ and } A \neq \begin{pmatrix} 1 & 0 \\ 0 & 1 \end{pmatrix} = 1_{M_2(\mathbb{R})}.$$

We can even have an infinitude of such elements. For example, if $A = \begin{pmatrix} 1 & x \\ 0 & 0 \end{pmatrix}$, $x \in \mathbb{R}$, then

$$A^2 = \begin{pmatrix} 1 & x \\ 0 & 0 \end{pmatrix} \begin{pmatrix} 1 & x \\ 0 & 0 \end{pmatrix} = \begin{pmatrix} 1 & x \\ 0 & 0 \end{pmatrix} = A.$$

Observe in passing that idempotent elements in rings are therefore not unique, unlike when considered in groups. However, if a is in some integral domain, then

$$a^2 = a \implies a^2 - a = 0 \implies a(a - 1) = 0 \implies a = 0 \text{ or } a = 1.$$

In other words, in an integral domain, the only idempotent elements are 0 and 1.

(12) True. By properties of a ring, we may write

$$a + b - ab = 1 \implies a + b - ab - 1 = 0 \implies (1 - a)b - (1 - a) = 0$$

so that $(1 - a)(1 - b) = 0$, from which we derive that either $a = 1$ or $b = 1$ because we are dealing with an integral domain.

(13) False again. It seems that we are going to stay in $(M_2(\mathbb{R}), +, \cdot)$ for a little while. For example, $A = \begin{pmatrix} 0 & 1 \\ 0 & 0 \end{pmatrix} \neq 0_{M_2(\mathbb{R})}$ and yet

$$A^2 = \begin{pmatrix} 0 & 0 \\ 0 & 0 \end{pmatrix} = 0_{M_2(\mathbb{R})}.$$

There are other examples. For example in $(\mathbb{Z}_4, +, \cdot)$, it is seen that

$$\overline{2}^2 = \overline{4} = \overline{0},$$

yet $\overline{2} \neq \overline{0}$.

Remark. In an integral domain, $a^2 = 0$ if and only if $a = 0$.

(14) False. In $(M_2(\mathbb{R}), +, \cdot)$, let $A = \begin{pmatrix} 0 & 1 \\ 0 & 0 \end{pmatrix}$ and $B = \begin{pmatrix} 1 & 0 \\ 0 & 0 \end{pmatrix}$. Then

$$AB = \begin{pmatrix} 0 & 0 \\ 0 & 0 \end{pmatrix} = 0_{M_2(\mathbb{R})}$$

whilst $BA = \begin{pmatrix} 0 & 1 \\ 0 & 0 \end{pmatrix} \neq 0_{M_2(\mathbb{R})}$.

(15) The implication "$\Rightarrow$" is true, and "$\Leftarrow$" is false. For a proof of the former case, observe that upon left multiplying by a and utilizing the associativity of "$\cdot$", one can write

$$ab = ba \implies a(ab) = a(ba) \implies a^2b = aba = (ab)a \implies a^2b = (ba)a = ba^2.$$

Remark. Similarly, it may be shown using a proof by induction that $ab = ba$ implies $ab^n = b^n a$ for all $n \in \mathbb{N}$, and the previous holds for all $n \in \mathbb{Z}$ when b is invertible. This remark will be used at some point below.

For a counterexample of the reverse implication, consider $A = \begin{pmatrix} 0 & 1 \\ 1 & 0 \end{pmatrix}$ and $B = \begin{pmatrix} 1 & 0 \\ 0 & 2 \end{pmatrix}$ as elements of $(M_2(\mathbb{R}), +, \cdot)$.

Since $A^2 = \begin{pmatrix} 1 & 0 \\ 0 & 1 \end{pmatrix} = I_2$, obviously $A^2 B = BA^2$. However,

$$AB = \begin{pmatrix} 0 & 2 \\ 1 & 0 \end{pmatrix} \neq \begin{pmatrix} 0 & 1 \\ 2 & 0 \end{pmatrix} = BA.$$

(16) False! First, we are not in say $(\mathbb{R}, +, \cdot)$. In other words, "+" and "$\cdot$" here are just some notations, so students should be wary of the emotional connotations of "+" and "$\cdot$". The expression $2x = 0$ means that $x + x = 0$, and so $x = -x$ only. To elaborate more, if $*$ is the first binary operation of a ring and e is the identity element with respect to $*$, then $2x = 0$ may be expressed as $x * x = e$. So, does the latter equation always give $x = e$? Still not convinced? Here is an explicit counterexample: In $\mathbb{Z}_6$, we have

$$2 \times \bar{3} = \bar{3} + \bar{3} = \bar{6} = \bar{0}$$

but $\bar{3} \neq \bar{0}$.

(17) The answer depends on char R! When char $R > 2$, it is true, and when char $R = 2$, it is false. So, we give a counterexample to the latter, and a proof of the former. In $\mathbb{Z}_2 = \{\bar{0}, \bar{1}\}$, the sum of its elements is $\bar{0} + \bar{1} = \bar{1} \neq \bar{0}$.

When char $R > 2$, each element is paired with its (additive) inverse, and given that the addition is commutative in a ring, it becomes clear that the sum of all elements is zero.

(18) False! In the ring $(M_2(\mathbb{R}), +, \cdot)$, take $A = \begin{pmatrix} 1 & 0 \\ 0 & 1 \end{pmatrix}$ and $B = -A$. Then A and B are both invertible (with respect to the multiplication of matrices). But $A + B = \begin{pmatrix} 0 & 0 \\ 0 & 0 \end{pmatrix}$ is non-invertible.

(19) False. While a subring is definitely a ring, it is not always true that the bigger ring "bequeathes" its identity element to one of its subrings. For a simple counterexample, remember that $(\mathbb{Z}, +, \cdot)$ is a (commutative) ring, but its subring $2\mathbb{Z}$ (see Exercise 5.3.8 for a proof) does not have an identity. Indeed, $2\mathbb{Z}$ possessing and identity $1_{2\mathbb{Z}}$ means that

$$1_{2\mathbb{Z}} \cdot x = x \cdot 1_{2\mathbb{Z}} = x$$

for all *even* integers x. So the only possible choice would be $1_{2\mathbb{Z}} = 1$ which is patently outside $2\mathbb{Z}$.

(20) True. This actually follows from the properties of group homomorphisms. Let us supply a proof anyway. Observe that

$0_R = 0_R + 0_R$, and so

$$f(0_R) = f(0_R + 0_R) = f(0_R) + f(0_R)$$
$$\implies f(0_R) - f(0_R) = f(0_R) + f(0_R) - f(0_R)$$

Hence

$$0_{R'} = f(0_R) + 0_{R'} = f(0_R)$$

as $f(0_R)$ is in R'.

(21) True. The proof is pretty simple. Let $x \in R$, then observe that

$$0 = f(0) = f[x + (-x)] = f(x) + f(-x)$$
$$\implies 0 - f(x) = -f(x) + f(x) + f(-x)$$

so that

$$-f(x) = 0 + f(-x) = f(-x),$$

as needed.

(22) False. We give two counterexamples. For example, let R be any ring and set $f(x) = 0$ for all $x \in R$. Then for all $x, y \in R$, $f(x) = f(y) = 0$ and $x + y, x \cdot y \in R$. Hence

$$f(x + y) = 0 = f(x) + f(y) \text{ and } f(x \cdot y) = 0 = f(x) \cdot f(y).$$

So, f is a ring homomorphism. Observe in the end that

$$f(1) = 0 \neq 1.$$

Another counterexample reads: Let $f : (\mathbb{R}, +, \cdot) \to (M_2(\mathbb{R}), +, \cdot)$ be defined by

$$f(x) = \begin{pmatrix} x & 0 \\ 0 & 0 \end{pmatrix}$$

for all $x \in \mathbb{R}$. Each of the two "+" and each of the two "·" has a different meaning here.

Then for all $x, y \in \mathbb{R}$

$$f(x + y) = \begin{pmatrix} x + y & 0 \\ 0 & 0 \end{pmatrix} = f(x) + f(y) \text{ and}$$

$$f(x \cdot y) = \begin{pmatrix} xy & 0 \\ 0 & 0 \end{pmatrix} = f(x) \cdot f(y).$$

This says that f is a ring homomorphism and yet

$$f(1) = \begin{pmatrix} 1 & 0 \\ 0 & 0 \end{pmatrix} \neq \begin{pmatrix} 1 & 0 \\ 0 & 1 \end{pmatrix} = 1_{M_2(\mathbb{R})}.$$

(23) True. Indeed, if $f(1) = 0$, then it follows that $f(-1) = 0$ (as $f(-x) = -f(x)$ for any ring homomorphism f). But, this would mean that f is not injective, and so we must have $f(1) \neq 0$. As $f(1) \neq 0$, it is invertible as S is a field, and so

$$f(1) = f(1 \cdot 1) = f(1) \cdot f(1) \text{ or } f(1) \cdot [f(1)]^{-1} = f(1) \cdot f(1) \cdot [f(1)]^{-1},$$

which gives $f(1) = 1$, as wished.

5.3. Exercises with Solutions

Exercise 5.3.1. Let $(R, +, \cdot)$ be some ring, and let $x, y, z \in R$. Show that the following statements are true:

(1) If $x + z = y + z$, then $x = y$.
(2) If $x + x = x$, then $x = 0$.
(3) $x \cdot 0 = 0 \cdot x = 0$.
(4) $-(-x) = x$.
(5) $-x \cdot y = x \cdot (-y) = (-x) \cdot y$.

Solution 5.3.1. At first sight, some students could wonder why a proof would be needed for such obvious statements anyway?! What deepens their confusion a bit more is the fact that they are still not used to the signs "+" and "$\cdot$" in a different environment from that of the usual numbers. In fact, even on $\mathbb{R}$, one has to show that, e.g. $x \times 0 = 0$, as this property is not one of the axioms of real numbers. Notice in the end of this paragraph that this is the kind of exercises that one has to do at least once in a life time.

(1) This statement follows easily from Exercise 4.3.1.
(2) Clearly, $x + x = x$ implies $x + x + (-x) = x + (-x)$, where $-x$ is the (additive) inverse of x with respect to "+". Hence $x + 0 = 0$, where 0 is the identity element still with respect to "+". For the same reason, one gets in the end $x = 0$.
(3) Since 0 is the identity element for "+", we have $0 = 0 + 0$. So, according to the distributivity of "$\cdot$" over "+", we get

$$x \cdot 0 = x \cdot (0 + 0) = x \cdot 0 + x \cdot 0.$$

If $-(x \cdot 0)$ designates the inverse of $x \cdot 0$ (with respect to "+"), then we obtain

$$0 = x \cdot 0 - x \cdot 0 = x \cdot 0 + x \cdot 0 - x \cdot 0 = x \cdot 0 + 0 = x \cdot 0.$$

The way of showing $0 \cdot x = 0$ is similar, hence we leave it to interested readers.

(4) A way of seeing why this is true is to remember that the inverse of the inverse of an element is this element.

(5) We give the proof without detailing each step. We may write

$$x[y + (-y)] = x \cdot 0 = 0 \implies xy + x(-y) = 0 \implies x(-y) = -xy.$$

Finally, readers may analogously show the other equality.

Exercise 5.3.2. Is $(\mathbb{Z}, -, \cdot)$ a ring with identity?

Solution 5.3.2. Since "$-$" is not associative over $\mathbb{Z}$, $(\mathbb{Z}, -)$ is not even a group. So, $(\mathbb{Z}, -, \cdot)$ cannot be a ring.

Exercise 5.3.3. We define two binary operations $*$ and $\perp$ over $\mathbb{R}$ by:

$$x * y = x + y - 1 \text{ and } x \perp y = x + y - xy.$$

Is $(\mathbb{R}, *, \perp)$ a ring?

Solution 5.3.3. The answer is positive, i.e. $(\mathbb{R}, *, \perp)$ a ring. First, we need to see why $(\mathbb{R}, *)$ is an abelian group. Since "$+$" is commutative in $\mathbb{R}$,

$$\forall x, y \in \mathbb{R}: \ x + y - 1 = y + x - 1.$$

In other words,

$$\forall x, y \in \mathbb{R}: \ x * y = y * x,$$

i.e. "$*$" is commutative in $\mathbb{R}$. The operation "$*$" is associative because

$$\forall x, y, z \in \mathbb{R}: \ x * (y * z) = (x * y) * z \ (= x + y + z - 2).$$

Let the interested reader check details.

Since "$*$" is commutative, to find the identity element, which we note e, we solve the equation $x * e = x$. Since

$$x * e = x + e - 1 = x,$$

it follows that $e = 1$ is the identity element. The inverse of $x \in \mathbb{R}$, with respect to "$*$", is $x' = 2 - x$. Thus, $(\mathbb{R}, *)$ is a commutative group.

Now, the binary operation $\perp$ is associative because

$$\forall x, y, z \in \mathbb{R}: \ (x \perp y) \perp z = x \perp (y \perp z),$$

which readers should either show that directly, or they just have to remember that this was already done in Exercise 4.3.42. It also saves time to observe that $\perp$ is commutative by Exercise 4.3.42 (remember that commutativity of the second operation is not compulsory).

Since $\perp$ is commutative, to show that $\perp$ is distributive over $*$, we need only check that

$$\forall x, y, z \in \mathbb{R}: \ x \perp (y * z) = (x \perp y) * (x \perp z).$$

Let $x, y, z \in \mathbb{R}$. Then

$$
\begin{aligned}
x \perp (y * z) &= x \perp (y + z - 1) \\
&= x + y + z - 1 - x(y + z - 1) \\
&= x + y + z - 1 - xy - xz + x \\
&= (x + y - xy) + (x + z - xz) - 1 \\
&= (x \perp y) * (x \perp z).
\end{aligned}
$$

Finally, observe that

$$
0 \perp x = 0 + x - 0 = x + 0 - 0 = x \perp 0 = x,
$$

for all real x. Since the identity element is unique, the previous equations say that 0 is the identity element with respect to $\perp$. Thus, $(\mathbb{R}, *, \perp)$ is a (commutative) ring with identity.

Exercise 5.3.4. We define on $\mathbb{R}^2$ two binary operations as:

$(x, y) + (x', y') = (x + x', y + y')$ and $(x, y) \bullet (x', y') = (xx', xy' + x'y)$.

Show that $(\mathbb{R}^2, +, \bullet)$ is a commutative ring with identity.

Solution 5.3.4. We already know that $(\mathbb{R}^2, +)$ is an abelian group, and that was Exercise 4.3.21.

As regards the second operation, observe that $\mathbb{R}^2$ is closed under "$\bullet$". That "$\bullet$" is associative is left to readers. The identity element with respect to $\bullet$ is $(1, 0)$. It is easy to see that $\bullet$ is commutative. So much for the axioms of the second operation.

There only remains to check the mixed axiom. But, since "$\bullet$" is commutative, to show the distributive law, it suffices to show that $\bullet$ is distributive on one side only with respect to $+$. To this end, let $(x, y), (x', y')$ and (x'', y'') all be in $\mathbb{R}^2$. Then

$$
\begin{aligned}
(x, y) \bullet [(x', y') + (x'', y'')] &= (x, y) \bullet (x' + x'', y' + y'') \\
&= (x(x' + x''), x(y' + y'') + (x' + x'')y)) \\
&= (xx' + xx'', xy' + xy'' + x'y + x''y) \\
&= (xx' + xx'', xy' + x'y + xy'' + x''y) \\
&= (xx', xy' + x'y) + (xx'', xy'' + x''y) \\
&= (x, y) \bullet (x', y') + (x, y) \bullet (x'', y''),
\end{aligned}
$$

and thus we have shown that $(\mathbb{R}^2, +, \bullet)$ is a commutative ring with identity, as required.

Exercise 5.3.5. Let $\mathcal{F}(\mathbb{R})$ be the set of functions defined from $\mathbb{R}$ into $\mathbb{R}$. Let $\circ$ be the usual composition of functions. Let $+$ be defined over $\mathcal{F}(\mathbb{R})$ as:

$$(f + g)(x) = f(x) + g(x), \ \forall x \in \mathbb{R}$$

for all f and g in $\mathcal{F}(\mathbb{R})$. Is $(\mathcal{F}(\mathbb{R}), +, \circ)$ a ring?

Solution 5.3.5. We already know from Exercise 4.3.13 that $(\mathcal{F}(\mathbb{R}), +)$ is a commutative group. However, while we have

$$\forall f, g, h \in E : \ (f + g) \circ h = (f \circ h) + (g \circ h),$$

we do not necessarily have

$$f \circ (g + h) = (f \circ g) + (f \circ h).$$

Indeed, let $f(x) = x - 1$, $g(x) = x$, and $h(x) = -x$, be all defined from $\mathbb{R}$ into $\mathbb{R}$. Then, for all $x \in \mathbb{R}$

$$(f \circ (g + h))(x) = f((g + h)(x)) = f(0) = -1$$

and

$$(f \circ g)(x) + (f \circ h)(x) = f(g(x)) + f(h(x)) = x - 1 - x - 1 = -2,$$

whereby

$$f \circ (g + h) \neq (f \circ g) + (f \circ h).$$

Thus, $(\mathcal{F}(\mathbb{R}), +, \circ)$ is not a ring.

Exercise 5.3.6. Let $\mathcal{F}(\mathbb{R})$ be the set of functions defined from $\mathbb{R}$ into $\mathbb{R}$. Let $\cdot$ and $+$ denote the usual pointwise multiplication and addition of functions respectively. Is $(\mathcal{F}(\mathbb{R}), +, \cdot)$ a ring?

Solution 5.3.6. The answer is positive, and $(\mathcal{F}(\mathbb{R}), +, \cdot)$ is a commutative ring with identity. Basic properties of $\mathbb{R}$, combined with Exercise 4.3.13 give a proof. Details are left to readers.

Remark. There is nothing special about functions from $\mathbb{R}$ into $\mathbb{R}$. In fact, and in a very similar manner, it may be shown that $(\mathcal{F}(I, \mathbb{R}), +, \cdot)$ is a ring, where I is any subset of $\mathbb{R}$.

Exercise 5.3.7. Let $M_2(\mathbb{R})$ be the set of 2×2 real matrices. Let "$+$" denote the usual addition of matrices, and let "$\cdot$" denote the multiplication of matrices. Show that $(M_2(\mathbb{R}), +, \cdot)$ is a ring. Is it commutative?

Solution 5.3.7. That $(M_2(\mathbb{R}), +)$ is a commutative group is already known (Exercise 4.3.16). The product of two matrices in $M_2(\mathbb{R})$ is obviously in $M_2(\mathbb{R})$. Other laws follow from Proposition 4.1.6. So, $(\mathcal{F}(\mathbb{R}), +, \cdot)$ is a ring.

In the end, the ring $(M_2(\mathbb{R}), +, \cdot)$ is not commutative as we already know that the product of two matrices need not be commutative.

Exercise 5.3.8. (Cf. Exercise 5.4.3) Let $n \in \mathbb{Z}$. Show that $n\mathbb{Z}$ is a subring of $(\mathbb{Z}, +, \cdot)$.

Solution 5.3.8. First, observe that $n\mathbb{Z}$ is not empty. Let $x, y \in n\mathbb{Z}$. Then, both x and y are multiples of n, as are x and $-y$. It becomes therefore evident that $x - y$ and xy remain multiples of n. In other words,

$$x - y, xy \in n\mathbb{Z},$$

and so $n\mathbb{Z}$ is indeed a subring of $(\mathbb{Z}, +, \cdot)$, as suggested.

Exercise 5.3.9. Let

$$\mathbb{Q}[\sqrt{2}] := \{x + y\sqrt{2} : x, y \in \mathbb{Q}\}.$$

Show that $(\mathbb{Q}[\sqrt{2}], +, \cdot)$ is a ring with identity.

Solution 5.3.9. We show that $\mathbb{Q}[\sqrt{2}]$ is a subring of the ring $(\mathbb{R}, +, \cdot)$. First, observe that $\mathbb{Q}[\sqrt{2}]$ is not empty as it contains $0 = 0 + 0\sqrt{2}$.

Now, let $a, a' \in \mathbb{Q}[\sqrt{2}]$, i.e. $a = x + y\sqrt{2}$ and $a' = x' + y'\sqrt{2}$, with $x, y, x', y' \in \mathbb{Q}$.

First, we show that $a - a' \in \mathbb{Q}[\sqrt{2}]$. We have

$$a - a' = x + y\sqrt{2} - x' - y'\sqrt{2}$$
$$= (x - x') + (y - y')\sqrt{2} \in \mathbb{Q}[\sqrt{2}],$$

because $x - x', y - y' \in \mathbb{Q}$, and we are half-way through.

Next, we show that $a \cdot a' \in \mathbb{Q}[\sqrt{2}]$. We have

$$a \cdot a' = (x + y\sqrt{2})(x' + y'\sqrt{2})$$
$$= (xx' + 4yy') + (xy' + x'y)\sqrt{2} \in \mathbb{Q}[\sqrt{2}]$$

for $(xx' + 4yy'), (xy' + x'y) \in \mathbb{Q}$. By the subring test, $\mathbb{Q}[\sqrt{2}]$ is a subring of the ring $(\mathbb{R}, +, \cdot)$.

Finally, the identity element is $1 + 0\sqrt{2}$, as it may be checked.

Exercise 5.3.10. Let I be any subset of $\mathbb{R}$ and let $C(I, \mathbb{R})$ be the set of real-valued *continuous* functions defined on I. Let "$\cdot$" and "$+$" denote the usual pointwise multiplication and addition of functions respectively. Is $(C(I, \mathbb{R}), +, \cdot)$ a ring?

Solution 5.3.10. The answer is positive. To see why, observe that $(C(I, \mathbb{R})$ is a subset of $\mathcal{F}(I, \mathbb{R})$, and since $(\mathcal{F}(I, \mathbb{R}), +, \cdot)$ is a ring (see the remark below Solution 5.3.6), it suffices to show that $(C(I, \mathbb{R}), +, \cdot)$ is a subring of it.

It is easy to see that $C(I, \mathbb{R})$ is not empty. For instance, it contains the zero function over I, and it is a continuous function. Now, let f, g be two real-valued continuous functions on I, i.e. $f, g \in C(I, \mathbb{R})$. Then $f - g$ and fg remain real-valued and defined on I. Basic properties of continuous functions also inform us that $f - g$ and fg stay continuous. Thus, we have shown that $(C(I, \mathbb{R}), +, \cdot)$ is a subring of $\mathcal{F}(I, \mathbb{R})$, and so it is itself a ring.

Exercise 5.3.11. Show that the set S constituted of matrices of the form $\begin{pmatrix} 0 & a \\ 0 & 0 \end{pmatrix}$, $a \in \mathbb{R}$, is a subring of $(M_2(\mathbb{R}), +, \cdot)$.

Solution 5.3.11. The set S is obviously non-void as the identity element of $(M_2(\mathbb{R}), +)$, i.e. the zero matrix, does belong to S (just take $a = 0$). Now, let $A = \begin{pmatrix} 0 & a \\ 0 & 0 \end{pmatrix} \in S$ and $B = \begin{pmatrix} 0 & b \\ 0 & 0 \end{pmatrix} \in S$, where $a, b \in \mathbb{R}$. The inverse of B with respect to "$+$" is $\begin{pmatrix} 0 & -b \\ 0 & 0 \end{pmatrix}$. Therefore,

$$A - B = \begin{pmatrix} 0 & a - b \\ 0 & 0 \end{pmatrix} \in S \text{ and } AB = \begin{pmatrix} 0 & 0 \\ 0 & 0 \end{pmatrix} \in S.$$

Thus, S is a subring of $(M_2(\mathbb{R}), +, \cdot)$.

Exercise 5.3.12. Call a real 2×2 matrix $A = \begin{pmatrix} a & b \\ c & d \end{pmatrix}$ symmetric if it is equal to $\begin{pmatrix} a & c \\ b & d \end{pmatrix}$ (the latter is known as the transpose of A, and it is designated by A^t). Is the set of symmetric matrices a subring of $(M_2(\mathbb{R}), +, \cdot)$?

Solution 5.3.12. The answer is negative. For instance, if we consider $A = \begin{pmatrix} 0 & 1 \\ 1 & 0 \end{pmatrix}$ and $B = \begin{pmatrix} 2 & 0 \\ 0 & 1 \end{pmatrix}$, then both A and B are symmetric. Nevertheless,

$$AB = \begin{pmatrix} 0 & 1 \\ 2 & 0 \end{pmatrix}$$

is non-symmetric.

Remark. It is noteworthy that the sum of two symmetric matrices stays symmetric.

Exercise 5.3.13. Let $(R, +, \cdot)$ be a ring. The center of R is defined by
$$Z(R) = \{x \in R : xr = rx \text{ for all } r \in R\}.$$
Show that $Z(R)$ is a subring of R.

Solution 5.3.13. We use the subring test. First, $Z(R)$ is non-empty because it contains 0 as $0 \cdot r = r \cdot 0 \, (= 0)$ for all r. Next, let $x, y \in Z(R)$. Let $r \in R$. By calculational rules and the laws for rings, we may write
$$(x - y)r = xr - yr = rx - ry = r(x - y).$$
So, $x - y \in Z(R)$. Finally, since
$$(xy)r = x(yr) = x(ry) = (xr)y = (rx)y = r(xy),$$
it ensues that $xy \in Z(R)$. Thus, $Z(R)$ is a subring of R.

Remark. When R is commutative, then $Z(R) = R$.

Exercise 5.3.14. Let R be a ring such that for all x in R: $x^2 = x$ (such a ring is more commonly known as a "Boolean ring").
(1) Show that $\forall x \in R : x = -x$. **Hint:** Evaluate $(x + x)^2$.
(2) Show that R is commutative. **Hint:** Evaluate $(x + y)^2$.
(3) Define a relation $\mathcal{R}$ over R by
$$\forall (x, y) \in R^2 : x\mathcal{R}y \iff xy = x.$$

Is $\mathcal{R}$ an order relation?

Solution 5.3.14. Denote the first binary operation by "$+$", and the second one by "$\cdot$".

(1) Let $x \in R$. Hence $x + x \in R$, and so
$$x + x = (x + x)^2 = (x + x)(x + x) = x^2 + x^2 + x^2 + x^2 = x + x + x + x,$$
since $x^2 = x$. Therefore,
$$x + \underbrace{x + (-x)}_{0} + (-x) = x + x + x + \underbrace{x + (-x)}_{0} + (-x)$$
but
$$x + 0 + (-x) = x + x + x + 0 + (-x)$$
$$\implies x + (-x) = x + x + \underbrace{x + (-x)}_{0} = x + x + 0.$$
Finally,
$$x + x = 0 \text{ or } x + \underbrace{x + (-x)}_{0} = 0 + (-x) = -x,$$
that is
$$x = -x.$$

(2) Let $x, y \in R$. We have to prove that $xy = yx$ (because "·" is the second operation). Since $x, y \in R$, $x^2 = x$ and $y^2 = y$. Arguing as in the previous answer gives

$$x + y = (x + y)^2 \implies x + y = x^2 + xy + yx + y^2$$
$$\implies x + y = x + xy + yx + y$$
$$\implies xy + yx = 0$$
$$\implies xy = -yx.$$

Since $yx \in R$, we know that $-yx = yx$ (by the first question) and so $xy = yx$, as wished.

(3) $\mathcal{R}$ is reflexive because $\forall x \in R : xx = x^2 = x$ by the definition of this ring R. It is also an anti-symmetric relation. Indeed,

$$\forall x, y \in R : \quad x\mathcal{R}y \text{ and } y\mathcal{R}x \iff \forall x, y \in R : \quad xy = x \text{ and } yx = y.$$

Since R is commutative, we get $xy = yx$, which yields $x = y$.

To show that $\mathcal{R}$ is transitive, let $x, y, z \in R$ and assume $x\mathcal{R}y$ and $y\mathcal{R}z$, i.e. $xy = x$ and $yz = y$. Then

$$x = xy = xyz = xz, \text{ i.e. } x\mathcal{R}z.$$

Therefore, we have shown that $\mathcal{R}$ is an order relation, as suggested.

Exercise 5.3.15. ([6]) Let $(R, +, \cdot)$ be a ring.

(1) Show that if $x^2 - x \in Z(R)$ for all $x \in R$, then R is commutative (where $Z(R)$ designates the center of R).

(2) Infer that if $x^2 + x \in Z(R)$ for all $x \in R$, then R is commutative.

Solution 5.3.15. ([6])

(1) Let $x, y \in R$. Then $x + y \in R$, and so $(x+y)^2 - (x+y) \in Z(R)$. But

$$(x+y)^2 - (x+y) = x^2 + xy + yx + y^2 - x - y = (x^2 - x) + (y^2 - y) + xy + yx$$

thanks to the commutativity of "+". Since $Z(R)$ is a (sub)ring, it follows that $xy + yx \in Z(R)$. Hence $x(xy + yx) = (xy + yx)x$ or $x^2 y + xyx = xyx + yx^2$. Therefore, $x^2 y = yx^2$, and so $x^2 \in Z(R)$. Thus, $x^2 - (x^2 - x) = x^2 - x^2 + x = x \in Z(R)$ for all x. This means that $Z(R) = R$. In other words, R is commutative, as requested.

(2) Since $x^2 + x \in Z(R)$ is valid for all $x \in R$, it is true in particular for $-x \in R$. That is, $(-x)^2 - x = x^2 - x \in R$, still valid for all x. By the preceding answer, we know that R must be commutative.

Exercise 5.3.16. ([6]) Let $(R, +, \cdot)$ be a ring and let $Z(R)$ be its center.

(1) Assume that in R we have the property that $x^2 = 0$ implies $x = 0$. Show that if $e \in R$ is such that $e^2 = e$ (recall that such an element is called idempotent), then $e \in Z(R)$.
(2) Deduce that if $x^3 = x$ for all $x \in R$, then $x^2 \in Z(R)$.
(3) Show that if for all $x \in R$: $x^3 = x$, then R is commutative.

Solution 5.3.16. ([6])

(1) Let $t \in R$. Then

$$(et - ete)^2 = etet + eteete - etete - eteet = etet + etete - etete - etet = 0.$$

In very much the same way, $(te - ete)^2 = 0$ is obtained. By assumption, we get

$$et - ete = te - ete = 0$$

for all $t \in R$. Hence $et = te$ for all t, i.e. $e \in Z(R)$, as wished.
(2) Suppose $x^2 = 0$ for a certain $x \in R$. Then

$$x = x^3 = x^2 \cdot x = 0 \cdot x = 0,$$

and so $x^2 = 0$ yields $x = 0$. Since

$$(x^2)^2 = x^4 = x^3 \cdot x = x \cdot x = x^2,$$

it is seen that x^2 is idempotent, and by the previous question $x^2 \in Z(R)$.
(3) In Reference [6], the writers provided plenty of proofs of the commutativity of the ring R (in which $x^3 = x$ for all x). Here is one of them: Let $x \in R$. Since $x^2 + x \in R$, it follows that $(x^2 + x)^3 = x^2 + x$, and so $(x^2 + x)^2 \in Z(R)$ by the previous question. In other words, and as $x^3 = x$,

$$(x^2 + x)^2 = x^4 + 2x^3 + x^2 = x^2 + 2x + x^2 = 2(x^2 + x) \in Z(R).$$

Moreover,

$$x^2 + x = (x^2 + x)^3 = x^6 + 3x^5 + 3x^4 + x^3 = x^2 + 3x + 3x^2 + x = 4(x^2 + x),$$

and so $3(x^2 + x) = 0 \in Z(R)$. Accordingly,

$$x^2 + x = 3(x^2 + x) - 2(x^2 + x) \in Z(R).$$

Statement (2) of Exercise 5.3.15 now intervenes and gives the commutativity of R, as looked forward to.

Exercise 5.3.17. Let $(R, +_R, \times_R)$ and $(S, +_S, \times_S)$ be two rings. There is an obvious way for defining two binary operations, denoted by "+" and "·", on the direct (or cartesian) product $R \times S$: If $x, x' \in R$ and $y, y' \in S$, then define

$$(x, y) + (x', y') = (x +_R x', y +_S y') \text{ and } (x, y) \cdot (x', y') = (x \times_R x', y \times_S y').$$

As is customary, "+" here is called the componentwise addition, whereas "·" is referred to as the componentwise multiplication.

Show that $(R \times S, +, \cdot)$ is a ring, with identity iff R and S possess identities. Show that R and S are commutative if and only if $R \times S$ is commutative.

Solution 5.3.17. The way of proving each of the statements is similar to that of Exercise 4.3.38. For example, $(R \times S, +)$ is already an abelian group by the same exercise. It is quite easy to see that "·" is associative and that the distributive laws hold, and all that is left to readers.

If the identities of R and S are denoted by 1_R and 1_S respectively, then $(1_R, 1_S)$ is the identity element of $R \times S$ as

$$(x, y) \cdot (1_R, 1_S) = (x \times_R 1_R, y \times_S 1_S) = (x, y)$$

for all $(x, y) \in R \times S$, and $(1_R, 1_S) \cdot (x, y) = (x, y)$ for all x and y as well. Conversely, if $(1, 1)$ is the identity element of $R \times S$, then

$$(1, 1) \cdot (x, y) = (x, y) \cdot (1, 1) = (x, y)$$

whichever $x \in R$ and $y \in S$, from which we derive $1 \times_R x = x$, i.e. the first coordinate of the identity element of $R \times S$ is the identity element of R. Similarly, we see that the second component of the identity element of $R \times S$ is the identity element of S.

The statement about the commutativity of "·" is once more left to readers.

Exercise 5.3.18. Are the following rings integral domains:
1. $(\mathbb{R}, +, \times)$,
2. The set of functions defined from $\mathbb{R}$ into $\mathbb{R}$ with respect to $+$ and $\times$,
3. The set of continuous functions from $\mathbb{R}$ into $\mathbb{R}$ with respect to $+$ and $\times$,
4. $(\mathbb{Z}_4, +, \times)$,
5. $\mathbb{Z}^2$ as a direct product of the ring $(\mathbb{Z}, +, \cdot)$ with itself.
6. $(M_2(\mathbb{R}), +, \cdot)$?

Solution 5.3.18.

(1) $(\mathbb{R}, +, \times)$ is an integral domain. This follows from the known axioms of the real numbers.
(2) No! Consider the functions

$$f(x) = \begin{cases} 1, & x > 0, \\ 0, & x \leq 0, \end{cases} \text{ and } g(x) = \begin{cases} 0, & x > 0, \\ 2, & x \leq 0. \end{cases}$$

So f and g are both non-null, but their product fg is null because

$$(fg)(x) = \begin{cases} 1 \times 0, & x > 0, \\ 0 \times 2, & x \leq 0, \end{cases} = \begin{cases} 0, & x > 0, \\ 0, & x \leq 0, \end{cases} = 0,$$

for all $x \in \mathbb{R}$.
(3) Simply consider the continuous functions $f, g : \mathbb{R} \to \mathbb{R}$ defined by $f(x) = x + |x|$ and $g(x) = x - |x|$ respectively. Then none of f and g is null, and yet

$$f(x)g(x) = x^2 - |x|^2 = x^2 - x^2 = 0$$

for all $x, y \in \mathbb{R}$.
(4) Recall that $\mathbb{Z}_4 = \{\bar{0}, \bar{1}, \bar{2}, \bar{3}\}$. It is not an integral domain because for instance:

$$\bar{2} \times \bar{2} = \bar{4} = \bar{0}$$

yet $\bar{2} \neq \bar{0}$.
(5) Untrue. For instance, $(1, 0) \cdot (0, 1) = (0, 0)$.
(6) Consider

$$A = \begin{pmatrix} 0 & 1 \\ 0 & 0 \end{pmatrix} \text{ and } B = \begin{pmatrix} 1 & 0 \\ 0 & 0 \end{pmatrix}.$$

Then $A \neq 0_{M_2(\mathbb{R})}$ and $B \neq 0_{M_2(\mathbb{R})}$ but

$$AB = \begin{pmatrix} 0 & 0 \\ 0 & 0 \end{pmatrix}.$$

Remark. In those references requiring an integral domain to first be a commutative ring, we need not go that far to explain why $(M_2(\mathbb{R}), +, \cdot)$ is not an integral domain. Indeed, the observation that $(M_2(\mathbb{R}), +, \cdot)$ is not commutative amply suffices.

Exercise 5.3.19. Say that an element a of a ring R is nilpotent provided $a^n = 0$ for some integer $n \geq 1$. The smallest n such that $a^n = 0$ is called the index of nilpotence (or nilpotency).

Let R be a ring with identity, noted 1, and let $x, y \in R$ be nilpotent.

(1) Show that xy and $x + y$ are both nilpotent whenever $xy = yx$.

(2) Are $x + y$ and xy necessarily nilpotent if x and y are non-commutative?

(3) Show that $1 - x$ is invertible.

Solution 5.3.19. By assumption, there are $n, m \in \mathbb{N}$ such that $x^n = 0$ and $y^m = 0$.

(1) Since $xy = yx$, $(xy)^n = x^n y^n = 0 \times y^n = 0$ (cf. Exercise 4.3.27), i.e. xy is nilpotent.

To show that $x + y$ is nilpotent, the use of the binomial theorem seems unavoidable. We have

$$(x + y)^{n+m} = \sum_{k=0}^{n+m} \binom{n+m}{k} x^k y^{n+m-k}.$$

When $k \geq n$, $x^k = 0$ whereby $x^k y^{n+m-k} = 0$, and so

$$(x + y)^{n+m} = 0.$$

On the other hand, if $k < n$, then $n + m - k \geq m$ and so $y^{n+m-k} = 0$. Therefore, $(x + y)^{n+m} = 0$ in this case as well. The proof is then complete.

(2) This is not true anymore. For instance, consider the matrices

$$A = \begin{pmatrix} 0 & 1 \\ 0 & 0 \end{pmatrix} \text{ and } B = \begin{pmatrix} 0 & 0 \\ 1 & 0 \end{pmatrix}.$$

Then, both A and B are nilpotent (verify that $A^2 = B^2 = 0_{M_2(\mathbb{R})}$). However,

$$AB = \begin{pmatrix} 1 & 0 \\ 0 & 0 \end{pmatrix} \text{ and } A + B = \begin{pmatrix} 0 & 1 \\ 1 & 0 \end{pmatrix}.$$

So, neither AB nor $A + B$ is nilpotent. Indeed, it may be shown (by induction) that for all n:

$$(AB)^n = \begin{pmatrix} 1 & 0 \\ 0 & 0 \end{pmatrix}$$

and

$$(A + B)^{2n} = \begin{pmatrix} 1 & 0 \\ 0 & 1 \end{pmatrix}, (A + B)^{2n+1} = \begin{pmatrix} 0 & 1 \\ 1 & 0 \end{pmatrix}.$$

(3) Let $x^n = 0$. Using properties of a ring, it may easily be shown that (as in Exercise 1.3.15)

$$1 - x^n = (1 - x)(1 + x + \cdots + x^{n-1}),$$

i.e.

$$1 = (1 - x)(1 + x + \cdots + x^{n-1}).$$

Similarly, it is seen that

$$1 = (1 + x + \cdots + x^{n-1})(1 - x).$$

Therefore, $1 - x$ is invertible, with the inverse given by $1 + x + \cdots + x^{n-1}$.

Exercise 5.3.20. Let R be a ring and let $x, y \in R$ be such that xy is nilpotent. Show that yx is nilpotent too.

Solution 5.3.20. Since xy is nilpotent, we know that $(xy)^n = 0$ for a certain natural number n. Hence

$$(yx)^{n+1} = \underbrace{(yx)(yx)(yx)\cdots(yx)}_{n+1 \text{ times}} = y\underbrace{(xy)(xy)\cdots(xy)}_{n \text{ times}} x = y(xy)^n x = 0,$$

and so yx too is nilpotent.

Exercise 5.3.21. (Exponential of an element in a ring) Let $(R, +, \cdot)$ be a ring with identity, noted 1. Let $a \in R$ be a nilpotent element such that its index of nilpotence is $p + 1$. Define the exponential of a, symbolically e^a, by

$$e^a = 1 + a + \frac{a^2}{2!} + \cdots + \frac{a^p}{p!}.$$

(1) Give examples of the exponential of some elements in different rings.
(2) Show that for all $x, y \in R$ such that x and y are nilpotent and $xy = yx$, then

$$e^{x+y} = e^x e^y = e^y e^x.$$

(3) Deduce that e^x is invertible for any nilpotent x. What is $(e^x)^{-1}$?
(4) Give an example of two nilpotent non-commuting x and y such that $e^x e^y \neq e^y e^x$.

Remark. In general, exponentials are not defined for nilpotent elements only. We have chosen this way as this is the best we can do using the means at hand. Indeed, we do not have the mathematical tools at the level of this book to give the general definition of an exponential of non-nilpotent elements as this requires a bit of topology. Anyhow, this is a tasty foretaste, limited though, of the general exponential.

Solution 5.3.21.

(1) For example, let R be any ring with an identity 1. Then $e^0 = 1$, i.e. the exponential of the additive identity gives the multiplicative identity.

In $(\mathbb{Z}_8, +, \cdot)$, it is seen that $\bar{2}^3 = \bar{0}$. So

$$e^{\bar{2}} = \bar{1} + \bar{2} + \frac{\bar{2}^2}{2!} = \bar{1} + \bar{2} + \bar{2} = \bar{5}.$$

The last example is given in the ring $(M_2(\mathbb{R}), +, \cdot)$. Consider $A = \begin{pmatrix} 0 & 1 \\ 0 & 0 \end{pmatrix}$. Then $A^2 = 0_{M_2(\mathbb{R})}$ and so

$$e^A = I + A = \begin{pmatrix} 1 & 0 \\ 0 & 1 \end{pmatrix} + \begin{pmatrix} 0 & 1 \\ 0 & 0 \end{pmatrix} = \begin{pmatrix} 1 & 1 \\ 0 & 1 \end{pmatrix}.$$

(2) Since $xy = yx$ and both x and y are nilpotent, Exercise 5.3.19 guarantees the nilpotence of $x + y$, whereby we may compute e^{x+y} according to our definition. The remaining part of the proof is purely notational, using the binomial theorem (by a glance at Exercise 5.3.19 once again). We expect students to be capable of finishing the proof...

(3) Let x be nilpotent. So, $-x$ too is nilpotent. Now, since $e^0 = 1$ and x commutes with $-x$, it ensues that

$$1 = e^0 = e^{x-x} = e^x e^{-x} = e^{-x} e^x.$$

This tells us that e^x is invertible and that $(e^x)^{-1} = e^{-x}$.

(4) Let

$$A = \begin{pmatrix} 0 & 1 \\ 0 & 0 \end{pmatrix} \text{ and } B = \begin{pmatrix} 0 & 0 \\ 1 & 0 \end{pmatrix}.$$

Then A and B are both nilpotent elements of $(M_2(\mathbb{R}), +, \cdot)$. It can be verified that $AB \neq BA$. Besides,

$$e^A = I + A = \begin{pmatrix} 1 & 1 \\ 0 & 1 \end{pmatrix} \text{ and } e^B = I + B = \begin{pmatrix} 1 & 0 \\ 1 & 1 \end{pmatrix}.$$

Finally,

$$e^A e^B = \begin{pmatrix} 2 & * \\ * & * \end{pmatrix} \neq \begin{pmatrix} 1 & * \\ * & * \end{pmatrix} = e^B e^A.$$

Remark. We have used "$*$" instead of actual numbers here for the element in the first row and the first column of $e^A e^B$ is different from the entry of the same position in the matrix $e^B e^A$, and this is sufficient to declare that $e^A e^B \neq e^B e^A$.

Exercise 5.3.22. Supply an example of a finite ring which is non-commutative.

Solution 5.3.22. Most examples of rings we have met so far have been either "commutative and finite" or "infinite". So, to get the desired example, we must come up with a different type of examples. This example is inspired by one in the ring of matrices which is not commutative with respect to "·". However, this ring is infinite. To remedy the situation, we use a fairly standard trick. It consists of considering the set R constituted of 2×2 matrices whose coefficients are elements in $\mathbb{Z}_2$, i.e.

$$R = \left\{ \begin{pmatrix} \bar{a} & \bar{b} \\ \bar{c} & \bar{d} \end{pmatrix}, \text{ where } \bar{a}, \bar{b}, \bar{c}, \bar{d} = \bar{0} \text{ or } \bar{1} \right\},$$

equipped with the binary operations $\oplus$ and $\otimes$ defined by:

$$A \oplus B = \begin{pmatrix} \bar{a} + \bar{x} & \bar{b} + \bar{y} \\ \bar{c} + \bar{z} & \bar{d} + \bar{t} \end{pmatrix} \text{ and } A \otimes B = \begin{pmatrix} \bar{a}\,\bar{x} + \bar{b}\bar{z} & \bar{a}\,\bar{y} + \bar{b}\bar{t} \\ \bar{c}\,\bar{x} + \bar{d}\bar{z} & \bar{c}\,\bar{y} + \bar{d}\bar{t} \end{pmatrix},$$

where $A = \begin{pmatrix} \bar{a} & \bar{b} \\ \bar{c} & \bar{d} \end{pmatrix}$ and $B = \begin{pmatrix} \bar{x} & \bar{y} \\ \bar{z} & \bar{t} \end{pmatrix}$, and where the addition and multiplication of the entries is carried out as in the ring $\mathbb{Z}_2$.

Readers are invited to check that R is in effect a ring with respect to the two binary operations defined above. Moreover, R is finite as it has 16 elements. So, there only remains to show that R is not commutative. To this end, let $A = \begin{pmatrix} \bar{0} & \bar{1} \\ \bar{0} & \bar{0} \end{pmatrix}$ and $B = \begin{pmatrix} \bar{1} & \bar{0} \\ \bar{0} & \bar{0} \end{pmatrix}$. Then

$$A \otimes B = \begin{pmatrix} \bar{0} & \bar{0} \\ \bar{0} & \bar{0} \end{pmatrix} \neq \begin{pmatrix} \bar{0} & \bar{1} \\ \bar{0} & \bar{0} \end{pmatrix} = B \otimes A.$$

Exercise 5.3.23. (Cf. [8]) Is the sum of invertible elements in a finite ring either 0 or 1?

Solution 5.3.23. ([8]) The answer is negative. Consider the ring of matrices $M_2(\mathbb{Z}_2)$ with coefficients in $\mathbb{Z}_2$ (see Exercise 5.3.22). That $M_2(\mathbb{Z}_2)$ is a finite ring has already been explained. Take

$$A = \begin{pmatrix} \bar{1} & \bar{1} \\ \bar{0} & \bar{1} \end{pmatrix} \text{ and } I = \begin{pmatrix} \bar{1} & \bar{0} \\ \bar{0} & \bar{1} \end{pmatrix}.$$

Then both A and I are invertible with respect to "·". More precisely, and in this case, each matrix equals its inverse, i.e. $A^{-1} = A$ and $I^{-1} = I$. For example, for A, we have

$$A^2 = \begin{pmatrix} \bar{1} & \bar{1} \\ \bar{0} & \bar{1} \end{pmatrix} \begin{pmatrix} \bar{1} & \bar{1} \\ \bar{0} & \bar{1} \end{pmatrix} = \begin{pmatrix} \bar{1} & \bar{2} \\ \bar{0} & \bar{1} \end{pmatrix} = \begin{pmatrix} \bar{1} & \bar{0} \\ \bar{0} & \bar{1} \end{pmatrix} = I.$$

However, the sum

$$A + I = \begin{pmatrix} \bar{1} & \bar{1} \\ \bar{0} & \bar{1} \end{pmatrix} + \begin{pmatrix} \bar{1} & \bar{0} \\ \bar{0} & \bar{1} \end{pmatrix} = \begin{pmatrix} \bar{2} & \bar{1} \\ \bar{0} & \bar{2} \end{pmatrix} = \begin{pmatrix} \bar{0} & \bar{1} \\ \bar{0} & \bar{0} \end{pmatrix}$$

is neither $0_{M_2(\mathbb{Z}_2)}$ nor $1_{M_2(\mathbb{Z}_2)}$.

Exercise 5.3.24. Decide, giving reasons, whether each of the following maps are ring homomorphisms:

(1) $f : (\mathbb{Z}, +, \cdot) \to (\mathbb{Z}, +, \cdot)$, with $f(n) = 2n$.

(2) $f : (\mathbb{C}, +, \cdot) \to (\mathbb{C}, +, \cdot)$, with $f(z) = \bar{z}$.

(3) $f : (M_2(\mathbb{R}), +, \cdot) \to (\mathbb{R}, +, \times)$ where $f\left[\begin{pmatrix} a & b \\ c & d \end{pmatrix}\right] = a$.

Solution 5.3.24.

(1) While $f(n + m) = f(n) + f(m)$ for all n, m, the property $f(nm) = f(n)f(m)$ is not satisfied for all $n, m \in \mathbb{Z}$. For instance,

$$f(2 \cdot 3) = f(6) = 12 \text{ whereas } f(2) \cdot f(3) = 4 \times 6 = 24,$$

thereby $f(2 \cdot 3) \neq f(2) \cdot f(3)$. Therefore, f is not a ring homomorphism.

(2) Let $z_1, z_2 \in \mathbb{C}$. By well-known properties of complex numbers, it is seen that

$$f(z_1 + z_2) = \overline{z_1 + z_2} = \overline{z_1} + \overline{z_2} = f(z_1) + f(z_2)$$

and

$$f(z_1 \cdot z_2) = \overline{z_1 \cdot z_2} = \overline{z_1} \cdot \overline{z_2} = f(z_1) \cdot f(z_2).$$

In other words, f is a ring homomorphism.

(3) f is not a ring homomorphism. As in the first case, for all $A, B \in M_2(\mathbb{R})$: $f(A + B) = f(A) + f(B)$, as it may easily be checked. However, $f(AB) = f(A) \times f(B)$ is not satisfied for all $A, B \in M_2(\mathbb{R})$. For example, if

$$A = \begin{pmatrix} 2 & 1 \\ 0 & 0 \end{pmatrix} \text{ and } B = \begin{pmatrix} 3 & 0 \\ 1 & 0 \end{pmatrix},$$

then

$$AB = \begin{pmatrix} 7 & 0 \\ 0 & 0 \end{pmatrix}.$$

So,

$$f(AB) = 7 \neq 6 = 2 \times 3 = f(A) \times f(B).$$

Thus, f is not a ring homomorphism.

Exercise 5.3.25. Show that ring isomorphisms preserve idempotent elements, as well as nilpotent ones.

Solution 5.3.25. The idea of proof is the same for both statements. Let $f : (R, +, \cdot) \to (S, +, \cdot)$ be a ring isomorphism.

For the first case, let a be in R and such that $a^2 = a$. We have to show that $f(a)$ too is idempotent, i.e. $[f(a)]^2 = f(a)$. Once we have written both the assumption and the objective, the proof becomes right before our eyes. Indeed,

$$[f(a)]^2 = f(a)f(a) = f(a \cdot a) = f(a^2) = f(a).$$

Now, let a be a nilpotent element in R, i.e. $a^p = 0$ for some natural number p. Then

$$[f(a)]^p = \underbrace{f(a)f(a) \cdots f(a)}_{p \text{ times}} = f(\underbrace{a \cdot a \cdots a}_{p \text{ times}}) = f(a^p) = f(0) = 0,$$

i.e. $f(a)$ is nilpotent.

Exercise 5.3.26. Show that ring isomorphisms preserve integral domains.

Solution 5.3.26. Let $f : R \to S$ be a ring isomorphism. The aim is to show that S is an integral domain whenever R is one. So, assume R is an integral domain, then let $x, y \in S$ be such that $xy = 0$. We need to obtain either $x = 0$ or $y = 0$.

Since $x \in S$ and f is onto, $f(a) = x$ for some $a \in R$. Similarly, $f(b) = y$ for a certain $b \in R$. So

$$xy = 0 \iff f(a)f(b) = 0 \iff f(ab) = 0.$$

Applying the inverse of f yields $ab = 0$. Since R is an integral domain, we must have either $a = 0$ or $b = 0$. Whence $f(a) = f(0) = 0$ or $f(b) = 0$. In other words, $x = 0$ or $y = 0$, as looked forward to.

Exercise 5.3.27. Can $(2\mathbb{Z}, +, \cdot)$ be isomorphic to $(3\mathbb{Z}, +, \cdot)$?

Solution 5.3.27. The answer is no! In order to get a contradiction, let us imagine that there is a ring isomorphism $f : (2\mathbb{Z}, +, \cdot) \to (3\mathbb{Z}, +, \cdot)$. Clearly, $f(2) = 3n$ for a certain integer n. Now, we use the definition of a ring homomorphism to find $f(4)$. On the one hand,

$$f(4) = f(2 + 2) = f(2) + f(2) = 3n + 3n = 6n,$$

and on the other hand

$$f(4) = f(2 \cdot 2) = f(2) \cdot f(2) = (3n) \cdot (3n) = 9n^2.$$

Hence $6n = 9n^2$ (as an equation on $\mathbb{Z}$). This equation has two solutions, namely: $n = 0$ and $n = 2/3$. Only the former solution is accepted.

Whence $f(2) = 0$, and we already know that $f(0) = 0$ (for any ring homomorphism). Thus, f cannot even be injective, thereby it cannot be bijective either.

Exercise 5.3.28. Can $(\mathbb{Z}_4, +, \cdot)$ be isomorphic to $(\mathbb{Z}_2 \times \mathbb{Z}_2, +, \cdot)$?

Solution 5.3.28. False again. There is a simple way of seeing this, alas, not within the scope of this manuscript. Thankfully, there is another way to present to students at this stage. Assume a ring isomorphism $f : (\mathbb{Z}_4, +, \cdot) \to (\mathbb{Z}_2 \times \mathbb{Z}_2, +, \cdot)$ does exist. So, $f(x) + f(x) = 0$ for all $x \in \mathbb{Z}_4$ (where it should be $\bar{x}$ instead of x) because for any $y \in \mathbb{Z}_2 \times \mathbb{Z}_2$, $y + y = 0$. Hence $f(1) + f(1) = (0, 0)$.
Since f is a homomorphism, it ensues that

$$f(2) = f(1 + 1) = f(1) + f(1) = (0, 0).$$

But $f(0) = (0, 0)$. As above, the last two conditions prevents f from injective.

Exercise 5.3.29. Show that the only ring isomorphism from $(\mathbb{Z}, +, \cdot)$ onto $(\mathbb{Z}, +, \cdot)$ is the identity map.

Solution 5.3.29. First, it is plain that the identity map is a ring isomorphism. Conversely, let $f : (\mathbb{Z}, +, \cdot) \to (\mathbb{Z}, +, \cdot)$ be a ring isomorphism, and we ought to show that $f(n) = n$ for all $n \in \mathbb{Z}$.
Assume for now that $n \geq 0$. Then $n = 1 + 1 + \cdots + 1$ (n 1s) and so

$$f(n) = f(1 + 1 + \cdots + 1) = f(1) + f(1) + \cdots + f(1) = nf(1)$$

(rigorously speaking, one should in fact perform a proof by induction). We are done with the case $n \geq 0$ if we come to show that $f(1) = 1$. This is pretty easy to see as

$$f(1) = f(1 \cdot 1) = f(1) \cdot f(1) \implies 1 = f(1)$$

as $f(1) \in \mathbb{Z}$ and $f(1) \neq 0$ (otherwise $f(1) = 0$ would mean that f is not one-to-one). Therefore, $f(n) = n$ for all $n \geq 0$.
To treat the case $n < 0$ (hence $-n > 0$), remember that $f(-x) = -f(x)$ for any $x \in \mathbb{Z}$. So, by this property and the first part of the proof, we have

$$-f(n) = f(-n) = -n,$$

and so $f(n) = n$ for negative n as well. Accordingly, $f(n) = n$ for all $n \in \mathbb{Z}$, as required.

Exercise 5.3.30. (Cf. Exercise 5.4.13) Show that any field is an integral domain.

Solution 5.3.30. To see why, assume that $(F, +, \cdot)$ is a field. Let $x, y \in F$ be such that $x \cdot y = 0$. For the sake of contradiction, assume that $x \neq 0$ and $y \neq 0$. Then x^{-1} exists as F is a field, and so

$$0 = x^{-1} \cdot 0 = x^{-1} \cdot x \cdot y = (x^{-1} \cdot x) \cdot y = 1 \cdot y = y,$$

which is the sought contradiction. Therefore, either $x = 0$ or $y = 0$, and this completes the proof.

Exercise 5.3.31. Show that any finite integral domain is a field.

Solution 5.3.31. Recall that our definition of an integral domain does not require the second binary operation to be commutative. Now, assume that $(R, +, \cdot)$ is an integral domain such that $\operatorname{card} R$ is finite, and let a be a non-zero element of R. We must show that a is invertible.

To this end, define two maps from R into R by

$$f_a(x) = a \cdot x \text{ and } g_a(x) = x \cdot a.$$

Clearly, both f_a and g_a are one-to-one. For example, let $x, y \in R$ and assume that $f_a(x) = f_a(y)$, i.e. $a \cdot x = a \cdot y$. Then

$$a \cdot (x - y) = a \cdot [x + (-y)] = 0.$$

Since $x - y = x + (-y) \in R$, and R is an integral domain and $a \neq 0$, we get $x - y = 0$ or $x = y$, establishing the injectivity of f_a. Similarly, readers may show that g_a is injective.

By Exercise 2.3.24 and the finiteness of R, both f_a and g_a are surjective. Since $1 \in R$, for some $c \in R$, $a \cdot c = 1$. Also, for some $d \in R$, $d \cdot a = 1$. By Exercise 4.3.24, we have that a is invertible, thus showing that R is a field, as required.

Exercise 5.3.32. Let $\mathbb{R}[X]$ be the set of polynomials with real coefficients.
 (1) Show that $(\mathbb{R}[X], +, \cdot)$ is a ring with identity.
 (2) Is it a field?

Remark. The addition and the multiplication of polynomials are already known to readers. They are carried out naturally. The addition is done by combining like terms, e.g.

$$(x^3 + 5x^2 - 4x) + (2x^4 - 2x^3 + 6x - 9)$$
$$= 2x^4 + (x^3 - 2x^3) + 5x^2 + (-4x + 6x) - 9$$
$$= 2x^4 - x^3 + 5x^2 + 2x - 9.$$

As for multiplication, for example

$$(2x^3 + 6x + 1) \cdot (x^5 + 3x^2) = 2x^8 + 6x^5 + 6x^6 + 18x^3 + x^5 + 3x^2,$$

so that

$$(2x^3 + 6x + 1) \cdot (x^5 + 3x^2) = 2x^8 + 6x^6 + 7x^5 + 18x^3 + 3x^2,$$

and this is the multiplication of the two polynomials $p(x) = 2x^3 + 6x + 1$ and $q(x) = x^5 + 3x^2$

Solution 5.3.32.

(1) We may go through all axioms occurring in the definition of a ring. Alternatively, and as $\mathbb{R}[X]$ is a subset of $\mathcal{F}(\mathbb{R})$ (the set of functions defined from $\mathbb{R}$ into $\mathbb{R}$), it suffices to show that $(\mathbb{R}[X], +, \cdot)$ is a subring of $(\mathcal{F}(\mathbb{R}), +, \cdot)$ (the latter is a ring by Exercise 5.3.6). First, observe that $\mathbb{R}[X]$ is non-void, then let p, q be two polynomials. Basic properties of polynomials allow us to see that $p - q, p \cdot q \in \mathbb{R}[X]$, whereby showing that $(\mathbb{R}[X], +, \cdot)$ is a ring. Finally, the identity element is the constant polynomial equal to 1.

(2) $(\mathbb{R}[X], +, \cdot)$ fails to be a field. The reason is that if we take a non-zero polynomial, e.g. $p(x) = x$, then $p \cdot q = 1$, where 1 is the constant polynomial taking the value 1 only, would entail $p(x)q(x) = 1$ for all $x \in \mathbb{R}$. That is, $xq(x) = 1$ for all $x \in \mathbb{R}$. In particular, when $x = 0$, we would obtain $0 = 1$ which is a contradiction. Since we have found a non-zero polynomial without an inverse, $(\mathbb{R}[X], +, \cdot)$ cannot be a field.

Exercise 5.3.33. Let $\mathbb{Z}[X]$ be the set of polynomials with integer coefficients. Show that $(\mathbb{Z}[X], +, \cdot)$ is an integral domain.

Solution 5.3.33. The way of showing that $(\mathbb{Z}[X], +, \cdot)$ is a ring with identity is similar to that of Answer 5.3.32.

Let neither $p(x)$ nor $q(x)$ be the zero polynomial. Assume that $p(x)$ is of degree n, and $q(x)$ is of degree m. The associated coefficients with n and m are denoted by a_n and a_m respectively. So, $a_n \neq 0$ and $a_m \neq 0$. Since $a_n, a_m \in \mathbb{Z}$ and $\mathbb{Z}$ is an integral domain, it ensues that $a_n a_m \neq 0$. But the coefficient $a_n a_m$ is associated with x^{n+m} which is the term of highest degree in the product $p(x)q(x)$. Therefore, $p(x)q(x)$ is not the zero polynomial. This shows that $(\mathbb{Z}[X], +, \cdot)$ is an integral domain, as needed.

Exercise 5.3.34. Let $n \in \mathbb{N}$. Show that $(\mathbb{Z}_n, \oplus, \otimes)$ is a field if and only if n is a prime number.

Solution 5.3.34. Well, this is an extremely important result in abstract algebra. However, there is not much left to show here. Indeed, using the equivalent definition of a field stated in Definition 5.1.5, and Exercises 4.3.10 & 4.3.12, we need only check the distributive laws.

These are inherited from the distributivity laws of $\mathbb{Z}$. Besides, and due to the commutativity of $\otimes$, it suffices to check either of the distributive laws. For all $\overline{x}, \overline{y}, \overline{z} \in \mathbb{Z}_n$, one has

$$\overline{x} \otimes (\overline{y} \oplus \overline{z}) = \overline{x} \otimes (\overline{y+z}) = \overline{x \cdot (y+z)} = \overline{x \cdot y + x \cdot z} = \overline{x \cdot y} \oplus \overline{x \cdot z},$$

and so

$$\overline{x} \otimes (\overline{y} \oplus \overline{z}) = (x \otimes y) \oplus (x \otimes z),$$

as wished.

Exercise 5.3.35. Let F be a finite field. Find $\prod_{x \in F^*} x$, i.e. the product of all non-zero element in F.

Solution 5.3.35. Each $x \neq 0$ in F is invertible. So, x and x^{-1} are both elements of F. The product of all such elements is -1 when $x \neq x^{-1}$. However, when $x = x^{-1}$, i.e. when $x^2 = 1$, then this equation implies that $x = 1$ or $x = -1$ as F is a field (hence an integral domain). Therefore, the product of all non-zero elements is -1 (if $1 \neq -1$).

Even in the event that $1 = -1$ (e.g. in $\mathbb{Z}_2$), the product is still $-1 = 1$. To recap, and in all cases, the product of non-zero elements is -1.

Exercise 5.3.36. Can the field $(\mathbb{R}, +, \cdot)$ be isomorphic to the field $(\mathbb{Q}, +, \cdot)$?

Solution 5.3.36. No! Assume there there exists a field isomorphism $f : (\mathbb{R}, +, \cdot) \to (\mathbb{Q}, +, \cdot)$. We do know that $f(1) = 1$ (cf. the "True or False" Section). Since f is a ring isomorphism, there are two ways here of writing $f(2)$. First,

$$f(2) = f(1 + 1) = f(1) + f(1) = 1 + 1 = 2.$$

Second,

$$f(2) = f(\sqrt{2} \cdot \sqrt{2}) = f(\sqrt{2})f(\sqrt{2}) = [f(\sqrt{2})]^2.$$

So, $[f(\sqrt{2})]^2 = 2$. By the definition of f, $f(\sqrt{2})$ is a rational number, and so the last equation does not have solutions in $\mathbb{Q}$. Thus, $(\mathbb{R}, +, \cdot)$ and $(\mathbb{Q}, +, \cdot)$ cannot be isomorphic.

Exercise 5.3.37. Let F and F' be two fields, and let $f : F \to F'$ be a field homomorphism. Show that f is either injective or $f(x) = 0_{F'}$ for all $x \in F$.

Solution 5.3.37. Assume that f is not one-to-one, and we show that we must have $f(x) = 0_{F'}$ for all $x \in F$. Since f is not injective, there are $a, b \in F$, $a \neq b$ and $f(a) = f(b)$. Set $c = a - b$, and so $c \neq 0$. Hence c is invertible. Since f is a ring homomorphism,

$$f(c) = f(a - b) = f[a + (-b)] = f(a) + f(-b) = f(a) - f(b) = 0_{F'}.$$

Now, let x be an arbitrary element of F. We obtain

$$f(x) = f(x \cdot_F c^{-1} \cdot_F c) = f(x \cdot_F c^{-1}) \cdot_{F'} f(c) = f(x \cdot_F c^{-1}) \cdot_{F'} 0_{F'} = 0_{F'},$$

as suggested.

Remark. If we impose $f(1) = 1$ in the definition of a field homomorphism (as done by many), then the second case never occurs! Let us give a proof: Let $x, y \in R$ and assume $f(x) = f(y)$. Then (as above)

$$f(x) - f(y) = 0 \Longrightarrow f(x - y) = 0.$$

If $x \neq y$, i.e. $x - y \neq 0$, then $x - y$ is invertible. So

$$1 = f(1) = f[(x-y)(x-y)^{-1}] = f(x-y)f[(x-y)^{-1}] = 0 \cdot f[(x-y)^{-1}] = 0$$

and this is a contradiction. Thus, $x = y$, i.e. f is injective.

Exercise 5.3.38. Show that the direct product of two (non-trivial) fields need not be a field (with respect to the operations introduced in Exercise 5.3.17).

Solution 5.3.38. Let R and S be two fields. To see why $R \times S$ is not a field, we must exhibit a non-zero element of $R \times S$ not possessing an inverse. We claim that $(a, 0_S)$ is an instance of that, where $a \in R$ is non-zero. Saying that $(a, 0_S)$ has an inverse means that

$$1_{R \times S} = (1_R, 1_S) = (a, 0_S)(x, y) = (ax, 0_S)$$

for some $(x, y) \in R \times S$. This is impossible as $0_S \neq 1_S$.

Exercise 5.3.39. Prove that the only subfield of $(\mathbb{Q}, +, \times)$ is $\mathbb{Q}$ itself. *Recall that a subfield F' of a field $(F, +, \times)$ is a non-empty subset of F such that F' is a subgroup of $(F, +)$, and a subgroup of $(F \setminus \{0\}, \times)$.*

Solution 5.3.39. Let S be a subfield of $\mathbb{Q}$. Then $1 \in S$. So $\mathbb{N} \subset S$ because

$$\forall n \in \mathbb{N}: n = \underbrace{1 + 1 + \cdots + 1}_{n \text{ times}} \in S$$

Observe that also $0 \in S$.

Since S is a subgroup, $-n \in S$. Therefore, $\mathbb{Z} \subset S$.

Finally, let $x = \frac{p}{q}$ where $p \in \mathbb{Z}$ and $q \in \mathbb{Z}^*$. Since S is a field, it must contain $1/q$ (the inverse of q). Hence $p, q \in S$ and so $\frac{p}{q} \in S$. Finally

$$S \subset \mathbb{Q} \text{ and } \mathbb{Q} \subset S \Longrightarrow S = \mathbb{Q},$$

marking the end of the proof.

Exercise 5.3.40. Let $p \in \mathbb{N}$ be such that $\sqrt{p} \notin \mathbb{Q}$. Set

$$\mathbb{Q}[\sqrt{p}] = \{a + b\sqrt{p} : a, b \in \mathbb{Q}\}.$$

Show that $(\mathbb{Q}[\sqrt{p}], +, \cdot)$ is a field. How about $(\mathbb{Z}[\sqrt{p}], +, \cdot)$?

Solution 5.3.40. The method is fairly similar to that used in Exercise 5.3.9. Readers are asked to show that $(\mathbb{Q}[\sqrt{p}], +, \cdot)$ is a subring of $(\mathbb{R}, +, \cdot)$. So $(\mathbb{Q}[\sqrt{p}], +, \cdot)$ is a ring. It has an identity element which is $1 = 1 + 0\sqrt{p}$.

Next, let $x \neq 0$. We have to show that x is invertible. Since $x \in \mathbb{Q}[\sqrt{p}]$ and $x \neq 0$, $x = a + b\sqrt{p}$ for some $a, b \in \mathbb{Q}$. So (by observing that $a - b\sqrt{p}$ does not vanish either as $\sqrt{p} \notin \mathbb{Q}$)

$$\frac{1}{x} = \frac{1}{a + b\sqrt{p}} = \frac{a - b\sqrt{p}}{a^2 - b^2 p} = \frac{a}{a^2 - b^2 p} - \frac{b}{a^2 - b^2 p}\sqrt{p} \in \mathbb{Q}[\sqrt{p}]$$

as it is in the form of the elements of $\mathbb{Q}[\sqrt{p}]$ because $a/(a^2 - b^2 p)$ and $-b/(a^2 - b^2 p)$ are both in $\mathbb{Q}$.

Regarding the second question, $(\mathbb{Z}[\sqrt{p}], +, \cdot)$ is not a field. Let us exhibit a non-zero element without an inverse. Elements of $\mathbb{Z}[\sqrt{p}]$ are of the form $a + b\sqrt{p}$ where $a, b \in \mathbb{Z}$. Since $\sqrt{p} = 0 + 1\sqrt{p}$, it is seen that $\sqrt{p}$ is a non-zero element of $\mathbb{Z}[\sqrt{p}]$. If $\sqrt{p}$ had an inverse, we could write

$$\sqrt{p}(a + b\sqrt{p}) = 1 \implies a\sqrt{p} = 1 - bp.$$

Since $1 - bp$ is an integer, $a\sqrt{p}$ must be an integer and this is only possible when $a = 0$. Hence $1 - bp = 0$ and so $bp = 1$. This would mean that $b = p = 1$, and this is impossible. Thus, $(\mathbb{Z}[\sqrt{p}], +, \cdot)$ is not a field.

Exercise 5.3.41. Consider the fields $\mathbb{Q}[i\sqrt{2}]$ and $\mathbb{Q}[\sqrt{2}]$ both equipped with the usual addition and multiplication of ordinary numbers. Can they be isomorphic?

Solution 5.3.41. The answer is negative. To see that, assume there is in effect an isomorphism $f : \mathbb{Q}[i\sqrt{2}] \to \mathbb{Q}[\sqrt{2}]$. Then $f(1) = 1$ (you may re-consult the "True or False" Section of this chapter).

Now, write

$$1 = f(1) = f[(-1) \cdot (-1)] = f(-1) \cdot f(-1) = [f(-1)]^2.$$

Since $\mathbb{Q}[\sqrt{2}]$ is an integral domain, it ensues that either $f(-1) = 1$ or $f(-1) = -1$. Since f is injective, the case $f(-1) = 1$ is already eliminated. So, only $f(-1) = -1$ seems to be possible for the moment.

But,

$$-1 = -2 + 1 = (i\sqrt{2})^2 + 1 \Longrightarrow$$
$$-1 = f(-1) = f[(i\sqrt{2})^2] + f(1) = [f(i\sqrt{2})]^2 + 1$$

or $-2 = [f(i\sqrt{2})]^2$. Since $f(i\sqrt{2}) = a + b\sqrt{2}$ for some rational numbers a and b, it is fairly simple to see that $-2 = [f(i\sqrt{2})]^2$ would contradict the irrationality of $\sqrt{2}$. Thus, $\mathbb{Q}[i\sqrt{2}]$ and $\mathbb{Q}[\sqrt{2}]$ cannot be isomorphic.

Exercise 5.3.42. Let $(R, +, \cdot)$ be a ring. Say that $a \in R$ is a square root of $b \in R$ provided $a^2 = b$ (we must NOT write $\sqrt{b} = a$ in general!). This notion generalizes square roots of real or complex numbers.

(1) Provide examples of square roots in different rings.
(2) Give examples of rings in which some elements do not possess any square root.
(3) Provide examples of rings in which some elements have two square roots only.
(4) Give examples of rings in which certain elements have four square roots only.
(5) Provide an example of a ring in which a certain element has three square roots only.
(6) Supply an example of a ring in which some elements have infinitely many square roots.
(7) Show that in a commutative integral domain, each element cannot have more than two square roots.
(8) Is the number of non-zero square roots always even?
(9) Let a be a square root of b, where $a, b \in R$. Show that a is invertible if and only if b is so.

Solution 5.3.42.

(1) In $(\mathbb{Z}_4, +, \cdot)$ which is finite, it is easy to find square roots. We have

$$\bar{0} \cdot \bar{0} = \bar{0}, \ \bar{1} \cdot \bar{1} = \bar{1}, \ \bar{2} \cdot \bar{2} = \bar{4} = \bar{0}, \text{ and } \bar{3} \cdot \bar{3} = \bar{1}.$$

Observe that $\bar{0}$ has two square roots: $\bar{0}$ and $\bar{2}$, while $\bar{1}$ has two square roots: $\bar{1}$ and $\bar{3}$. Also, we remark that $\bar{2}$ and $\bar{3}$ do not possess any square root.

   Other examples have been given in the ring $(M_2(\mathbb{R}), +, \cdot)$ in the "True or False" Section. We may also add square roots in $\mathbb{R}$ or $\mathbb{C}$ which are well-known to readers.
(2) We have just seen that, e.g. $\bar{2}$ does not admit any square root in $(\mathbb{Z}_4, +, \cdot)$. Let us give other examples. In $(\mathbb{R}, +, \cdot)$, -1 does

not have any square root. Also, in $(\mathbb{Q}, +, \cdot)$, 2 does not have any square root.

We give one more example. In $(M_2(\mathbb{R}), +, \cdot)$. Let

$$A = \begin{pmatrix} 0 & 1 \\ 0 & 0 \end{pmatrix}.$$

Then A does not have any square root. To see this, suppose A has a square root, noted B, which is a 2×2 matrix of the form

$$B = \begin{pmatrix} a & b \\ c & d \end{pmatrix}.$$

Then

$$B^2 = A \Longleftrightarrow \begin{pmatrix} a^2 + bc & ab + bd \\ ac + cd & bc + d^2 \end{pmatrix} = \begin{pmatrix} 0 & 1 \\ 0 & 0 \end{pmatrix}.$$

So, we need to solve the system

$$\begin{cases} a^2 + bc = 0, \\ ab + bd = 1, \\ ac + cd = 0, \\ bc + d^2 = 0. \end{cases}$$

The equation $(a+d)c = ac+cd = 0$ (commutativity is available here!) gives either $c = 0$ or $a = -d$. If $c = 0$, the first and the fourth equations of the system become $a^2 = 0$ and $d^2 = 0$, whereby $a = d = 0$. But, this is not consistent with the second equation (look at $0 = 1$!). Therefore, $c = 0$ does not yield anything.

Let us now examine the case $a = -d$. In this case, it is seen that

$$ab + bd = 1 \Longleftrightarrow ab - ab = 1 \Longleftrightarrow 0 = 1,$$

which is impossible. Thus, the above system fails to have a solution, whereby the equation $B^2 = A$ is not satisfied by any matrix B. In other language, A does not have any square root.

(3) Technically, $\mathbb{R}$ or $\mathbb{C}$ are not excluded, and examples are readily available in these rings. Other examples were given in the first answer. Let us give another example. For example, the matrix $A = \begin{pmatrix} 1 & 0 \\ 0 & 0 \end{pmatrix}$ has two square roots only, given by

$$\begin{pmatrix} 1 & 0 \\ 0 & 0 \end{pmatrix} \text{ and } \begin{pmatrix} -1 & 0 \\ 0 & 0 \end{pmatrix}.$$

(4) In $(\mathbb{Z}_8, +, \cdot)$, $\bar{1}$ has four different square roots, namely: $\bar{1}$, $-\bar{1}$ $(=\bar{7})$, $\bar{3}$, and $-\bar{3}$ $(=\bar{5})$.

As another example, take $A = \begin{pmatrix} 1 & 0 \\ 0 & 2 \end{pmatrix}$ in $(M_2(\mathbb{R}), +, \cdot)$.

This is not a random choice, alas, no explanation can be revealed at this stage! Then A has four square roots only given by:

$$\begin{pmatrix} 1 & 0 \\ 0 & \sqrt{2} \end{pmatrix}, \begin{pmatrix} -1 & 0 \\ 0 & \sqrt{2} \end{pmatrix}, \begin{pmatrix} 1 & 0 \\ 0 & -\sqrt{2} \end{pmatrix} \text{ and } \begin{pmatrix} -1 & 0 \\ 0 & -\sqrt{2} \end{pmatrix}.$$

(5) Consider the ring $(\mathbb{Z}_9, +, \cdot)$. It may be checked that

$$\bar{0} \cdot \bar{0} = \bar{0}, \ \bar{3} \cdot \bar{3} = \bar{9} = \bar{0} \text{ and } \bar{6} \cdot \bar{6} = \bar{36} = \bar{0},$$

whereas for any $\bar{p} \neq \bar{0}, \bar{3}, \bar{6}$: $\bar{p}^2 \neq \bar{0}$. This says that $\bar{0}$ has exactly three different square roots.

(6) An efficient place to find such an example is once again the ring $(M_2(\mathbb{R}), +, \cdot)$. Notice that we have already observed in the "True or False" Section that $\begin{pmatrix} 1 & 0 \\ 0 & 1 \end{pmatrix}$ has an infinitude of square roots given by the family $\begin{pmatrix} x & 1 \\ 1 - x^2 & -x \end{pmatrix}$, $x \in \mathbb{R}$.

Another example over the same ring is the infinite family of matrices $\begin{pmatrix} 0 & x \\ 0 & 0 \end{pmatrix}$, $x \in \mathbb{R}$. For each real x, the latter is a square root of $0_{M_2(\mathbb{R})}$ for

$$\begin{pmatrix} 0 & x \\ 0 & 0 \end{pmatrix}^2 = \begin{pmatrix} 0 & 0 \\ 0 & 0 \end{pmatrix}.$$

(7) Let a and c be two elements in a commutative integral domain which are both square roots of some b, i.e. $a^2 = c^2 = b$. Hence $a^2 - c^2 = 0$, but

$$(a - c)(a + c) = a^2 - ca + ac - c^2 = a^2 - c^2$$

due to the commutativity of "$\cdot$". Thus, $(a - c)(a + c) = 0$, and so $a = c$ or $a = -c$ as we are in an integral domain.

Remark. In particular, in a commutative integral domain, the only square root of 0 is 0 itself.

(8) The answer depends on the type of the ring. For example, in $(\mathbb{Z}_3, +, \cdot)$, $\bar{1}$ has two different square roots, namely $\bar{2}$ and $\bar{1}$ as

$$\bar{2} \cdot \bar{2} = \bar{1} = \bar{1} \cdot \bar{1}.$$

(remember that may write $\bar{2} = -\bar{1}$).

On the other hand, in $(\mathbb{Z}_2, +, \cdot)$, e.g. $\bar{1}$ is the only square root of itself.

However, as long as we keep away from rings in which $a = -a$, i.e. those of characteristic 2 (e.g. $\mathbb{Z}_2$), the number of non-zero square roots is even. The proof is pretty obvious, and it reads: If a is a square root of a certain b, i.e. $a^2 = b$, then $-a$ too is a square root of b as

$$(-a)^2 = a^2 = b$$

by the usual calculational rules in rings.

(9) Let a and b be such that $a^2 = b$. If a is invertible, then so is a^2, by Proposition 4.1.1. Conversely, if b is invertible, then $bc = cb = 1$ for some $c \in R$. Since $a^2 = b$, it results that

$$a^2 c = ca^2 = 1 \iff a(ac) = (ca)a = 1.$$

Exercise 4.3.24 then implies that a is invertible, as wished.

Exercise 5.3.43. Solve the equations:
(1) $x^2 + x + \bar{1} = \bar{0}$ in $\mathbb{Z}_2$.
(2) $x^2 + x = \bar{0}$ in $\mathbb{Z}_5$.
(3) $x^2 + x = \bar{0}$ in $\mathbb{Z}_6$.
(4) $x^2 - \bar{2}x + \bar{1} = \bar{0}$ in $\mathbb{Z}_{11}$.
(5) $x^2 + x + \bar{5} = \bar{0}$ in $\mathbb{Z}_7$.
(6) $x^2 + x + \bar{5} = \bar{0}$ in $\mathbb{Z}_{17}$.
(7) $x^2 - \bar{4}x + \bar{3} = \bar{0}$ in $\mathbb{Z}_{12}$.

Solution 5.3.43.
(1) If $x = \bar{0}$, $x^2 + x + \bar{1} = \bar{1} \neq \bar{0}$, and if $x = \bar{1}$, $x^2 + x + \bar{1} = \bar{3} = \bar{1} \neq \bar{0}$. So, the given equation does not have any solution in $\mathbb{Z}_2$.
(2) The only solutions are $\bar{0}$ and $\bar{4}$. Indeed,

$$\bar{0}^2 + \bar{0} = \bar{0} \text{ and } \bar{4}^2 + \bar{4} = \overline{20} = \bar{0}$$

whilst

$$\bar{1}^2 + \bar{1} = \bar{2} \neq \bar{0}, \ \bar{2}^2 + \bar{2} = \bar{6} = \bar{1} \neq \bar{0} \text{ and } \bar{3}^2 + \bar{3} = \overline{12} = \bar{2} \neq \bar{0}.$$

(3) The solutions are $\bar{0}$, $\bar{2}$, $\bar{3}$ and $\bar{5}$. This is left to readers.
(4) Write

$$x^2 - \bar{2}x + \bar{1} = \bar{0} \iff (x - \bar{1})^2 = \bar{0}.$$

Since $\mathbb{Z}_{11}$ is an integral domain, we get $x = \bar{1}$ only.
(5) Since $\mathbb{Z}_7$ has only seven elements, we can check all of them. First, we have

$$\bar{0}^2 = \bar{0}, \ \bar{1}^2 = \bar{1}, \ \bar{2}^2 = \bar{4}, \ \bar{3}^2 = \bar{2}, \ \bar{4}^2 = \bar{2}, \ \bar{5}^2 = \bar{4} \text{ and } \bar{6}^2 = \bar{1}.$$

Then we calculate $x^2 + x + \bar{5}$ for each value of x in $\mathbb{Z}_7 = \{\bar{0}, \bar{1}, \cdots, \bar{6}\}$. Readers can therefore see that only $\bar{1}$ and $-\bar{2} = \bar{5}$ are solutions of the equation $x^2 + x + \bar{5} = \bar{0}$.

(6) The cardinal of $\mathbb{Z}_{17}$ is 17, so there are more cases to check. So, we adopt another method here. We carry over the method of the completion of squares to $\mathbb{Z}_n$. Since $\overline{18} = \bar{1}$, we can write

$$x^2 + x + \bar{5} = \bar{0} \Longleftrightarrow x^2 + \overline{18}x + \bar{5} = \bar{0} \Longleftrightarrow (x + \bar{9})^2 - \overline{76} = \bar{0}$$

or merely $(x + \bar{9})^2 = \overline{76} = \bar{8}$. If we compute all $\bar{p}^2$, $p \in \mathbb{Z}_{17}$, we see that only $\bar{5}^2 = \bar{8}$ and $\overline{12}^2 = \bar{8}$. So, what is the square root to consider? The following observation resolves this issue: $-\overline{12} = \bar{5}$ and $-\bar{5} = \overline{12}$.

So, the solutions are

$$x = \bar{5} - \bar{9} = -\bar{4} = \overline{13}, \ x = -\bar{5} - \bar{9} = -\overline{14} = \bar{3}.$$

Remark. It is worth noticing that since $\mathbb{Z}_{17}$ is a field, a quadratic equation cannot have more than two solutions (see Exercise 6.4.1 below). Since we have found two solutions, we know that there aren't any others. In other words, dealing with a field renders things much simpler to handle.

(7) We can write (can't we?)

$$\bar{0} = x^2 - \bar{4}x + \bar{3} = (x - \bar{3})(x - \bar{1}).$$

But, we cannot conclude directly that $x = \bar{1}$ or $x = \bar{3}$, and the reason is that $\mathbb{Z}_{14}$ is not an integral domain! So, how can we deal with such situation? It seems that the recourse left to us is brute force. So, we write instead of the above

$$x^2 - \bar{4}x + \bar{3} = 0 \Longleftrightarrow x^2 - \bar{4}x + \bar{4} - \bar{1} = \bar{0} \Longleftrightarrow (x - \bar{2})^2 = \bar{1}.$$

We need to know what the square roots of $\bar{1}$ are? Writing all squares of elements of $\mathbb{Z}_{12}$, it is seen that $\bar{1}^2 = \bar{1}$, $\bar{5}^2 = \bar{1}$, $\bar{7}^2 = \bar{1}$ and $\overline{11}^2 = \bar{1}$. Thus,

$$x - \bar{2} = \bar{1} \text{ or } x - \bar{2} = \bar{5} \text{ or } x - \bar{2} = \bar{7} \text{ or } x - \bar{2} = \overline{11}$$

or

$$x = \bar{3} \text{ or } x = \bar{7} \text{ or } x = \bar{9} \text{ or } x = \overline{13} = \bar{1}.$$

Remark. Alternatively, there is another way, quite similar though, which is to use the symbolic discriminant, denoted by $\bar{\Delta}$. We have

$$\bar{\Delta} = \overline{16} - \bar{4} \times \bar{3} = \overline{16} - \overline{12} = \bar{4}.$$

What are the square roots of $\overline{4}$ in $\mathbb{Z}_{12}$? Writing all squares of elements of $\mathbb{Z}_{12}$, it is seen that $\overline{2}^2 = \overline{4}$, $\overline{4}^2 = \overline{4}$, $\overline{8}^2 = \overline{4}$ and $\overline{10}^2 = \overline{4}$. If we write all possible "$(\overline{4} \pm \sqrt{\overline{\Delta}})/\overline{2}$", we see that there are numbers which do not satisfy the given equation, but we also find that $\overline{3}, \overline{7}, \overline{9}, \overline{1}$ are in effect solutions, as above.

Exercise 5.3.44. Let $(R, +, \cdot)$ be a ring. Let $a, b \in R$, where a is invertible. Assume that b commutes with both a^p and a^q, i.e. $ba^p = a^p b$ and $ba^q = a^q b$ for some relatively prime numbers p and q. Show that $ba = ab$, i.e. b commutes with a.

Remark. The previous exercise appeared in a much more advanced version (far beyond the level of students who use the present manuscript) in [**10**].

Solution 5.3.44. We have that $\alpha p + \beta q = 1$ for some integers α and β, by Bézout's theorem in arithmetic. Since a is an invertible element, so are a^p and a^q (we need invertibility as we will take negative powers of some elements). Since $ba^p = a^p b$ and $ba^q = a^q b$, it follows that $ba^{\alpha p} = a^{\alpha p} b$ and $ba^{\beta q} = a^{\beta q} b$ (see the remark below Answer 15 in the "True or False" Section of the present chapter). Hence

$$ba^{\alpha p} = a^{\alpha p} b \implies ba^{\alpha p} a^{\beta q} = a^{\alpha p} b a^{\beta q} = a^{\alpha p} a^{\beta q} b.$$

Therefore,

$$ba^{\alpha p + \beta q} = a^{\alpha p + \beta q} b \text{ or } ab = ba,$$

as required.

Exercise 5.3.45. Let p and q be two relatively prime numbers, and let $(F, +, \cdot)$ be a field. Let $a, b \in F$ be such that $a^p = b^p$ and $a^q = b^q$. Show that $a = b$.

Solution 5.3.45. If $a = 0$, then $b^p = 0$ necessarily leads to $b = 0$ (and so $a = b = 0$). Indeed, if $b \neq 0$, then b would be invertible, and so b^p would be invertible too. In particular, we would have $b^p \neq 0$. A similar reason applies if we assume $b = 0$, that is, we obtain $a = 0$.

So, assume that $a \neq 0$ *and* $b \neq 0$. Since p and q are co-prime numbers, Bezout's theorem says that $up + vq = 1$ for some integers u and v (only one of them is negative). WLOG, suppose that u is the negative integer. Now, $a^p = b^p$ yields $a^{up} = b^{up}$, and $a^q = b^q$ implies that $a^{vq} = b^{vq}$. Therefore, $a^{up} a^{vq} = b^{up} b^{vq}$

$$a = a^1 = a^{up+vq} = b^{up+vq} = b^1 = b,$$

as coveted.

5.4. Supplementary Exercises

Exercise 5.4.1. Let:
$$A = \{2a\sqrt{3} : a \in \mathbb{Z}\}.$$
Is $(A, +, \times)$ a ring?

Exercise 5.4.2. Give an example of a subring of a ring with identity, but the subring itself fails to have an identity element.

Exercise 5.4.3. Show that the only subrings of $\mathbb{Z}$ are the sets of the form $n\mathbb{Z}$ where $n \in \mathbb{Z}$.

Exercise 5.4.4. Is the intersection of two subrings of a ring R a subring of R? What about the union of two subrings?

Exercise 5.4.5. Let R be a ring and let $a \in R$. Show that $S := \{x \in R : ax = 0\}$ is a subring of R.

Exercise 5.4.6. Let R be a ring and let u be an idempotent element in R. Set $S = \{uau : a \in R\}$. Show that S is a ring with u as an identity.

Exercise 5.4.7. ([**6**]) Let R be a ring such that $(xy)^3 = xy$ for all $x, y \in R$. Show that R is commutative.

Exercise 5.4.8. ([**6**]) Let R be a ring such that $(xyz)^3 = xyz$ for all $x, y, z \in R$. Is R necessarily commutative?

Exercise 5.4.9. Let A be a ring such that for all $x \in A$: $x^4 = x$. Show that A is commutative.

Exercise 5.4.10. Let $(R, +, \cdot)$ be a ring. Say that $a \in R$ is a cube root of $b \in R$ provided $a^3 = b$.
  (1) Provide examples of cube roots in different rings.
  (2) (Cf. [**27**]) Give an example of an element having a square root but without any cube root.
  (3) (Cf. [**27**]) Give an example of an element having a cube root but without any square root.

Exercise 5.4.11. Is $(2\mathbb{Z}, +, \cdot)$ isomorphic to $(4\mathbb{Z}, +, \cdot)$?

Exercise 5.4.12. Let $(R, +, \cdot)$ be a ring. Show that the set
$$R' = \{(a, a) : a \in R\}$$
is a subring of $R \times R$. Is R' isomorphic to $R \times R$?

Exercise 5.4.13. (Cf. Exercise 5.3.30) Give an example of an integral domain which is not a field.

Exercise 5.4.14. Assume that R is a commutative ring of prime characteristic p. Prove that the freshman's dream holds, i.e.

$$(x + y)^p = x^p + y^p$$

for all $x, y \in R$.

Exercise 5.4.15. Let p be a prime number. Find the solutions of the equation $x^2 + x = \bar{0}$ in $\mathbb{Z}_p$.

Exercise 5.4.16. Let $p, q \in \mathbb{N}$ be such that $\sqrt{p}, \sqrt{q} \notin \mathbb{Q}$. Give a necessary and sufficient condition that makes the fields $\mathbb{Q}[\sqrt{p}]$ and $\mathbb{Q}[\sqrt{q}]$ isomorphic.

Exercise 5.4.17. Let F be a field. Show that char(F) is either 0 or a prime p.

CHAPTER 6

Polynomials and Rational Fractions

6.1. Basics

6.1.1. Basic definitions and operations.

We have already dealt with polynomials on several occasions. Let us recall the definition once more:

DEFINITION 6.1.1. A polynomial p is an expression defined by

$$p(x) = a_n x^n + a_{n-1} x^{n-1} + \cdots + a_1 x + a_0$$

where the coefficients $a_0, a_1, \cdots, a_{n-1}, a_n$ are complex numbers.

The degree of a polynomial $p(x)$, written $\deg p(x)$, is the highest power of x associated with a non-zero coefficient. The latter coefficient is said to be the leading coefficient. The constant polynomial is $p(x) = a_0$ (where we allow $a_0 = 0$ as well).

The set of polynomials with complex coefficients is denoted by $\mathbb{C}[X]$, and more generally by $F[X]$ when the coefficients belong to a field F.

Here we say that $p(x) = 0$ if and only if $a_i = 0$ for all $i = 0, 1, \cdots, n$. If this is the case, p is called the zero polynomial .

Remark. We may also denote polynomials by $P(X)$.

Remark. Notice that the definition of $p(x) = 0$ is different from when we are required to the solutions (s) of $p(x) = 0$. For example, $x^2 - 4x + 3$ is not the zero polynomial, but when we want to find roots of $x^2 - 4x + 3$, then we write $x^2 - 4x + 3 = 0$ (whose solutions are $x = 3$ or $x = 1$).

We have already defined the addition and the multiplication of polynomials in an informal way. The formal definitions are not needed in this book. See, e.g. [**7**].

Now, we say a word about polynomial division. The first result says that the usual division theorem for integers may be carried over to polynomials (see, e.g. [**32**] for a proof):

THEOREM 6.1.1. *(Euclidean division) Assume we are given two polynomials $f(x)$ and $g(x)$ where $g(x) \neq 0$. Then there are unique* *polynomials $q(x)$ and $r(x)$ such that*

$$f(x) = g(x)q(x) + r(x),$$

where either $r(x) = 0$, or $r(x) \neq 0$ with $\deg r(x) < \deg g(x)$.

Remark. The polynomial $q(x)$ is called the quotient, and $r(x)$ is said to be the remainder.

Remark. The polynomial long division, which is to elaborated in the exercises' section, is an algorithm for Euclidean division.

DEFINITION 6.1.2. Let F be a field. Let $f(x), g(x) \in F[X]$ where $g(x) \neq 0$. Say that $g(x)$ divides $f(x)$ (or that B is a divisor of $f(x)$) provided that $f(x) = g(x)q(x)$ for some $q(x) \in F[X]$.

Next, we define the greatest common divisor of two polynomials.

DEFINITION 6.1.3. Let $f(x)$ and $g(x)$ be two polynomials. The greatest common divisor of $f(x)$ and $g(x)$, written $\gcd(f(x), g(x))$, is a polynomial of highest degree which divides both $f(x)$ and $g(x)$. We agree that the greatest common divisor is a polynomial whose leading coefficient equals 1.
When $\gcd(f(x), g(x)) = 1$, we say that the polynomials $f(x)$ and $g(x)$ are relatively prime (or co-prime).

The next notion is known to readers at least in the elementary context:

DEFINITION 6.1.4. Let F be a field. If $f(x) = a_n x^n + a_{n-1} x^{n-1} + \cdots + a_1 x + a_0 \in F[X]$ is a polynomial of degree n, and $c \in F$ is such that

$$f(c) = a_n c^n + a_{n-1} c^{n-1} + \cdots + a_1 c + a_0 = 0,$$

then c is called a root or a zero of $f(x)$.

THEOREM 6.1.2. *(Remainder theorem) Let F be a field, and let $f(x) \in F[X]$. If $c \in F$, then there is a polynomial $q(x)$ obeying*

$$f(x) = (x - c)q(x) + f(c).$$

In addition, $f(c) = 0$ if and only if $(x - c)$ divides $f(x)$.

Remark. In other words, the remainder of the division of $f(x)$ by $(x - c)$ is $f(c)$.

6.1.2. Factorization in $F[X]$, where F is a field.

Now, we pass to an important notion which is the factorization of polynomials in fields such as $\mathbb{Q}$, $\mathbb{R}$ or $\mathbb{C}$.

DEFINITION 6.1.5. Let F be a field and let $f(x) \in F[X]$ be a polynomial of degree at least one. Say that $f(x)$ is reducible in $F[X]$ provided there are polynomials $p(x), q(x) \in F[X]$ such that $f(x) = p(x)q(x)$, with $\deg p(x) \geq 1$ and $\deg q(x) \geq 1$. Otherwise, $f(x)$ is said to be irreducible.

THEOREM 6.1.3. *(The fundamental theorem of algebra, see, e.g. [17] for a proof). Every non-constant polynomial over $\mathbb{C}$ has a complex root. We could also say that $\mathbb{C}$ is algebraically closed.*

6.1.3. Rational fractions.

DEFINITION 6.1.6. A rational fraction (or function) is a fraction whose numerator and denominator are both polynomials.
The set of real rational fractions is denoted by $\mathbb{R}(X)$.

Next, we show how to factorize rational fractions (to obtain the so-called partial fraction decomposition). We assume that the numerator and the denominator are relatively prime.

- Let $f(x)/g(x)$ be a rational fraction. Then, there exist unique polynomials $q(x)$ and $r(x)$ such that

$$\frac{f(x)}{g(x)} = q(x) + \frac{r(x)}{g(x)},$$

  with $\deg r(x) < \deg g(x)$.
  The polynomials $q(x)$ and $g(x)$ are usually obtained via a Euclidean division.

- Let $f(x)/g(x)$ be a rational fraction such that $\deg f(x) < \deg g(x)$, and where $g(x) = (x-a)^n h(x)$ with $h(a) \neq 0$. Then $f(x)/g(x)$ may (uniquely) be decomposed as

$$\frac{f(x)}{g(x)} = \frac{\alpha_n}{(x-a)^n} + \frac{\alpha_{n-1}}{(x-a)^{n-1}} + \cdots + \frac{\alpha_1}{x-a} + \frac{k(x)}{h(x)}$$

  where $\alpha_n, \alpha_{n-1}, \cdots, \alpha_1 \in \mathbb{R}$, with $\deg k(x) < \deg h(x)$.

- In case $f(x)/g(x) \in \mathbb{R}(X)$ is such that $\deg f(x) < \deg g(x)$, and where $g(x) = (x^2+bx+c)^n h(x)$ with $b^2 - 4c < 0$ and $x^2 + bx+c$ and $h(x)$ being co-prime, then $f(x)/g(x)$ may (uniquely) be decomposed as

$$\frac{f(x)}{g(x)} = \frac{\alpha_n x + \beta_n}{(x^2+bx+c)^n} + \frac{\alpha_{n-1}x + \beta_{n-1}}{(x^2+bx+c)^{n-1}} + \cdots + \frac{\alpha_1 x + \beta_1}{x^2+bx+c} + \frac{k(x)}{h(x)}$$

where $\alpha_n, \alpha_{n-1}, \cdots, \alpha_1; \beta_n, \beta_{n-1}, \cdots, \beta_1 \in \mathbb{R}$, with $\deg k(x) <$ $\deg h(x)$.

Remark. Partial fraction decompositions over the complex numbers are not concerned with the third case of the previous procedure.

Remark. Partial fraction decomposition is a very useful tool for finding integrals (in calculus) of rational functions (see, e.g. [**26**]).

6.2. True or False

Questions. Determine, giving reasons, whether the following statements are true or false.

(1) The expressions $2x^2 + 3/x + 2$ and $x + 2\sqrt{x} - 8$ are polynomials.

(2) The exponential function $x \mapsto e^x$, defined from $\mathbb{R}$ into $\mathbb{R}$, is a polynomial.

(3) Let $f(x)$ and $g(x)$ be two polynomials such that $f(x) + g(x) \neq 0$. Then
$$\deg[f(x) + g(x)] = \max(\deg f(x), \deg g(x)).$$

(4) There is a polynomial $f(x)$ such that $[f(x)]^2 = x$.

(5) Irreducible polynomials in $\mathbb{R}[X]$ are those of degree one.

(6) Irreducible polynomials in $\mathbb{C}[X]$ are those of degree one.

(7) Let P be a polynomial of degree n, with coefficients in $\mathbb{C}$. Then P admits n roots (not necessarily distinct).

(8) We always know how factorize a polynomial into a product of irreducible elements.

(9) If z is a root of a polynomial in $\mathbb{C}$, then $\bar{z}$ too is a root of the same polynomial.

(10) Let $a, b, c \in \mathbb{C}$, with $a \neq 0$. If the solutions of $ax^2 + bx + c = 0$ form a pair of complex conjugates, then a, b and c must all be real.

(11) Every polynomial of odd degree in $\mathbb{R}$ has a root in $\mathbb{R}$.

(12) Every complex polynomial which is injective is necessarily of degree 1.

(13) If P is a complex polynomial such that $P(n) \in \mathbb{Z}$ for all $n \in \mathbb{Z}$, then the coefficients of P are all integers.

(14) Without doing any calculation, explain why
$$\frac{2}{x^2 + 1} + \frac{1}{x + 1} + \frac{1}{x - 1}$$
cannot be the partial fraction decomposition of
$$\frac{4x^2}{(x^2 - 1)(x^2 + 1)}.$$

(15) There is a rational fraction F such that $[F(x)]^2 = x$.

Answers.

(1) False! The first expression fails to be a polynomial as one of its terms has a negative power, namely: -1. The second expression, even if all powers are positive this time, one of them is not a natural number, viz. $1/2$.

(2) False! To show that $x \mapsto e^x$ is not a polynomial we use a contradiction argument. Suppose $x \mapsto e^x$ is a polynomial of degree n say, i.e. for certain coefficients $a_0, a_1, \cdots, a_n$ we have

$$e^x = a_0 + a_1 x + \cdots + a_n x^n.$$

Differentiating both sides of the last equation $(n+1)$-times yields

$$(e^x)^{(n+1)} = e^x = (a_0 + a_1 x + \cdots + a_n x^n)^{(n+1)} = 0,$$

and this is absurd. Accordingly, the exponential function is not a polynomial.

(3) False! Let us supply a counterexample. Take $f(x) = -x^2 + 2x$ and $g(x) = x^2$. Then $f(x) + g(x) \neq 0$ and $\deg f(x) = \deg g(x) = 2$, but

$$\deg[f(x) + g(x)] = \deg(2x) = 1 < 2 = \max(\deg f(x), \deg g(x)).$$

Notice that under the given assumptions of the question, we always have

$$\deg[f(x) + g(x)] \leq \max(\deg f(x), \deg g(x)).$$

(4) False! First, recall that if $f(x)$ and $g(x)$ are two non-zero polynomials, then

$$\deg[f(x)g(x)] = \deg f(x) + \deg g(x).$$

So, if we want $[f(x)]^2 = x$ for some polynomial $f(x)$ of degree n say, then we must have

$$1 = \deg(x) = \deg[f(x)]^2 = 2 \deg f(x) = 2n,$$

which is impossible as $n \in \mathbb{N}$.

(5) In fact, irreducible polynomials in $\mathbb{R}[X]$ are those of degree one, and those of degree two whose discriminant is strictly negative.

Remark. Any polynomial $f(x) \in \mathbb{R}[X]$ may uniquely (up to the order of the factors) be expressed as

$$f(x) = \alpha(x - a_1)^{n_1} \cdots (x - a_p)^{n_p}(x^2 + b_1 x + c_1)^{m_1} \cdots (x^2 + b_q x + c_q)^{m_q},$$

where $n_1, \cdots, n_p; m_1, \cdots, m_q \in \mathbb{N}$, and $\alpha, a_1, \cdots, a_p; b_1, \cdots, b_q$ and $c_1, \cdots, c_q$ are all real. Also, the discriminant of each of $x^2 + b_i x + c_i$ is negative.

(6) True! This a consequence of the fundamental theorem of algebra.

(7) True. This an equivalent formulation of the fundamental theorem of algebra stated above. Indeed, if P is a polynomial of degree n having a certain complex root z_1, then

$$P(z) = a(z - z_1)Q(z)$$

where Q is some polynomial of degree $n - 1$ and $a \neq 0$. Then Q too has a complex root. Proceeding as before as many times as needed, we obtain in the end that

$$P(z) = a(z - z_1) \cdots (z - z_n).$$

(8) False! According to the fundamental theorem of algebra, we "only" know that such a decomposition exists over the complex numbers.

Remark. Remember that Abel showed that equations of degree 5 and higher cannot be solved in radicals. The proof uses the so-called Galois theory (see, e.g. [**7**]). See [**28**] for an interesting paper about all that.

(9) False! Let $P(z) = z^2 + 2iz + 1$. Then $P(z)$ has complex coefficients, and has the following roots

$$z_1 = -(1 + \sqrt{2})i \text{ and } z_2 = -(1 - \sqrt{2})i,$$

where it is seen that z_1 is not the conjugate of z_2.

However, the result is true when the polynomial $P(x)$ solely real coefficients (see Exercise 6.3.1 for a proof):

THEOREM 6.2.1. *Let the polynomial $P(x)$ have real coefficients only, and assume that z_0 is a root of $P(x)$. Then $\overline{z_0}$ is also a root of $P(x)$.*

(10) False! Consider $x^2 + 2x + 3 = 0$. Then it has two complex solutions, namely $-1 - i\sqrt{2}$ and $-1 + i\sqrt{2}$, and it is obviously a pair of complex conjugates. However, $i(x^2 + 2x + 3) = 0$ too has the same pair of conjugate solutions, and yet $i, 2i, 3i$ are all complex numbers.

(11) True! See Exercise 6.3.2 for proofs.

(12) Let P be a complex polynomial. When $\deg P = 0$, i.e. when P is a constant polynomial, then P is not one-to-one. If $\deg P \geq 2$, then, given that P is a complex polynomial, two cases must be discussed.

(a) If P has two distinct roots, denoted by a and b say, then P is not injective for $P(a) = P(b) = 0$.

(b) Otherwise, $P(X) = \alpha(X-a)^n$ where $\alpha \in \mathbb{C}, \alpha \neq 0$ (this is possible as P is defined on $\mathbb{C}$). But then $P(X) - \alpha$ would admit n distinct roots, given by $a + \lambda$ where λ represents any of the n-th roots of unity. Indeed,

$$P(X) - \alpha = \alpha(X - a)^n - \alpha = \alpha[(X - a)^n - 1] = 0$$

yields $(X - a)^n = 1$ because $\alpha \neq 0$. So, $P(X)$ is not injective.

(13) False! Consider

$$P(X) = \frac{1}{2}X^2 + \frac{1}{2}X = \frac{X(X+1)}{2}.$$

For all $n \in \mathbb{Z}$, $P(n) \in \mathbb{Z}$ as $n(n+1)$ is always even. But the coefficients of P are rationals.

(14) The answer is simple. The rational fraction $\frac{4x^2}{(x^2-1)(x^2+1)}$ is even whilst the suggested partial fraction decomposition is not even. Readers can check that the correct partial fraction decomposition is:

$$\frac{2}{x^2 + 1} - \frac{1}{x + 1} + \frac{1}{x - 1}.$$

(15) False! Assume $F(x) = p(x)/q(x)$ where $p(x)$ and $q(x)$ are co-prime polynomials. If $[F(x)]^2 = x$, then

$$[p(x)]^2 = x[q(x)]^2 \implies \deg[p(x)]^2 = \deg(x[q(x)]^2)$$
$$\implies 2\deg p(x) = \deg(x) + 2\deg q(x)$$
$$\implies 2[\deg p(x) - \deg q(x)] = \deg(x) = 1,$$

and this is absurd for the very left side is an even number.

6.3. Exercises with Solutions

Exercise 6.3.1. Let the polynomial $P(x)$ have real coefficients only, and assume that z_0 is a root of $P(x)$. Show that $\overline{z_0}$ is also a root of $P(x)$.

Solution 6.3.1. Let $P(x) = a_n x^n + a_{n-1} x^{n-1} + \cdots + a_1 x + a_0$, where $a_i \in \mathbb{R}$ for all $i = 0, 1, \cdots, n$. Since z_0 is a root of $P(x)$, $P(z_0) = 0$, i.e.

$$a_n z_0^n + a_{n-1} z_0^{n-1} + \cdots + a_1 z_0 + a_0 = 0.$$

Hence

$$\overline{a_n z_0^n + a_{n-1} z_0^{n-1} + \cdots + a_1 z_0 + a_0} = \overline{0} = 0.$$

Since $\overline{z^n} = \overline{z}^n$ for all z, and by other basic properties of complex numbers, we know that the last displayed equation may be re-written as

$$\overline{a_n}\, \overline{z_0}^n + \overline{a_{n-1}}\, \overline{z_0}^{n-1} + \cdots + \overline{a_1}\, \overline{z_0} + \overline{a_0} = 0.$$

Since $a_i \in \mathbb{R}$ for all $i = 0, 1, \cdots, n$, the previous equation becomes

$$a_n \overline{z_0}^n + a_{n-1} \overline{z_0}^{n-1} + \cdots + a_1 \overline{z_0} + a_0 = 0.$$

In other words, $P(\overline{z_0}) = 0$, i.e. $\overline{z_0}$ is a root of $P(x)$, as suggested.

Exercise 6.3.2. Show that any polynomial of odd degree over $\mathbb{R}$ has a root in $\mathbb{R}$.

Solution 6.3.2. We give two proofs.

(1) First proof: Let $P(x)$ be a real polynomial of odd degree, say $2n + 1$, $n \in \mathbb{N}$. By the fundamental theorem of algebra, $P(x)$ has $2n + 1$ roots (distinct or not). By Exercise 6.3.1, these roots occur in conjugate pairs. Since the number of roots is odd, one root must be equal to its complex conjugate, i.e. this root has to be real, as needed.

(2) Second proof (see [**30**] for a more rigorous and detailed proof): The proof uses the intermediate value theorem (from basic real analysis). Let $P(x) = a_n x^n + \cdots + a_1 x + a_0$ where n is odd and $a_n \neq 0$. If $a_n > 0$, then for very large positive x, $P(x)$ is positive, and it is negative for very large negative x. Since P is continuous, it must pass through the real axis. An akin reasoning applies when $a_n < 0$.

Exercise 6.3.3. Decide which of the following polynomials are irreducible in $F[X]$:

(1) $2X + 3$; $F[X] = \mathbb{Q}[X], \mathbb{R}[X], \mathbb{C}[X]$,
(2) $X^2 - 4X + 3$; $F[X] = \mathbb{Q}[X], \mathbb{R}[X], \mathbb{C}[X]$,
(3) $X^2 - 2$; $F[X] = \mathbb{Q}[X], \mathbb{R}[X], \mathbb{C}[X]$,
(4) $X^2 + 4$; $F[X] = \mathbb{Q}[X], \mathbb{R}[X], \mathbb{C}[X]$.
(5) $P(X) = \overline{1}X^2 + \overline{1}$, $F[X] = \mathbb{Z}_3[X]$.

Solution 6.3.3.

(1) The polynomial $2X + 3$ is irreducible in all of $\mathbb{Q}[X]$, $\mathbb{R}[X]$ and $\mathbb{C}[X]$.
(2) The polynomial $X^2 - 4X + 3$ is reducible in all of $\mathbb{Q}[X]$, $\mathbb{R}[X]$ and $\mathbb{C}[X]$ for

$$X^2 - 4X + 3 = (X - 1)(X - 3).$$

(3) The polynomial $X^2 - 2$ is irreducible in $\mathbb{Q}[X]$. However, it is reducible in both $\mathbb{R}[X]$ and $\mathbb{C}[X]$ as

$$X^2 - 2 = (X - \sqrt{2})(X + \sqrt{2}).$$

(4) The polynomial $X^2 + 4$ is irreducible in $\mathbb{Q}[X]$ and in $\mathbb{R}[X]$, but it is reducible in $\mathbb{C}[X]$ since

$$X^2 + 1 = (X + i)(X - i).$$

(5) The given polynomial is irreducible in $\mathbb{Z}_3[X]$. Indeed, the possible roots in $\mathbb{Z}_3[X]$ are $\bar{0}$, $\bar{1}$ and $\bar{2}$. But

$$P(\bar{0}) = \bar{1}, P(\bar{1}) = \bar{2} \text{ and } P(\bar{2}) = \bar{5} = \bar{2}.$$

Thus, $P(X)$ has no roots in $\mathbb{Z}_3[X]$.

Exercise 6.3.4. Is the polynomial $X^4 + X^2 + 1$ reducible in $\mathbb{R}[X]$?

Solution 6.3.4. The answer is positive as its degree is greater than or equal to 3. To factorize it, we use complex numbers. Set $Y = X^2$, and so the given polynomial becomes $Y^2 + Y + 1$. The latter has two complex roots, namely $(-1 - i\sqrt{3})/2$ and $(-1 + i\sqrt{3})/2$. So in the polar form, we may then write

$$
\begin{aligned}
X^4 + X^2 + 1 &= (X^2 - e^{(2i\pi)/3})(X^2 - e^{(4i\pi)/3}) \\
&= (X - e^{(i\pi)/3})(X + e^{(i\pi)/3})(X - e^{(2i\pi)/3})(X + e^{(2i\pi)/3}) \\
&= (X - e^{(i\pi)/3})(X + e^{(2i\pi)/3})(X + e^{(i\pi)/3})(X - e^{(2i\pi)/3}) \\
&= (X^2 - X + 1)(X^2 + X + 1),
\end{aligned}
$$

and this is best we can do over the real numbers, because the discriminants of both factors in the last product are strictly negative.

Exercise 6.3.5. Is the polynomial $2X^3 - 2X + 1$ reducible in $\mathbb{Q}[X]$?

Solution 6.3.5. The answer is negative. Using the rational root test (Exercise 1.4.13), the possible rational roots of $P(X) := 2X^3 - 2X + 1$ are ± 1 and $\pm 1/2$. But none of these numbers is a root of $P(X)$, as

$$P(-1) = P(1) = 1, \ P(-1/2) = 7/4 \text{ and } P(1/2) = 1/4.$$

Since $P(X)$ has no roots over the rational numbers, $P(X)$ is irreducible in $\mathbb{Q}[X]$.

Exercise 6.3.6. Let F be a field, and let $f(x) = a_n x^n + a_{n-1} x^{n-1} + \cdots + a_1 x + a_0 \in F[X]$. The derivative of $f(x)$ is defined by

$$f'(x) = n a_n x^{n-1} + a_{n-1}(n-1) x^{n-2} + \cdots + a_1.$$

Let

$$P(X) = X^3 - X^2 - 8X + 12.$$

(1) Find $\gcd(P(X), P'(X))$.

(2) Factorize $P(X)$ into irreducible elements.

Solution 6.3.6.

(1) We have

$$P'(X) = 3X^2 - 2X - 8.$$

We use polynomial long division, but the notation could be unusual to some readers (they will get used to it quickly):

$$
\begin{array}{ll}
X^3 - X^2 - 8X + 12 & \big|\, 3X^2 - 2X - 8 \\
\underline{-(X^3 - \frac{2}{3}X - \frac{8}{3}X)} & \quad \frac{1}{3}X - \frac{1}{9} \\
-\frac{1}{3}X^2 - \frac{16}{3}X + 12 & \\
\underline{-(-\frac{1}{3}X^2 + \frac{2}{9}X + \frac{8}{9})} & \\
-\frac{50}{9}(X - 2) &
\end{array}
$$

Hence

$$X^3 - X^2 - 8X + 12 = (3X^2 - 2X - 8)\left(\frac{1}{3}X - \frac{1}{9}\right) - \frac{50}{9}(X - 2).$$

Dividing $3X^2 - 2X - 8$ by $X - 2$, we find $3X + 2$ and the remainder is null. The last non-zero remainder is $-\frac{50}{9}(X-2)$. The greatest common divisor is therefore

$$\gcd(P(X), P'(X)) = X - 2.$$

(2) Since $\gcd(P(X), P'(X)) = X - 2$, it is seen that $P(X)$ is divisible by $(X - 2)^2 = X^2 - 4X + 4$. Dividing $P(X)$ by $(X - 2)^2$ gives the following decomposition of $P(X)$:

$$P(X) = (X - 2)^2 (X + 3).$$

Exercise 6.3.7. Let $P(X) = X^4 - 4X^3 + 11X^2 - 14X + 10$.

(1) Check that $1 - i$ is a simple root of $P(X)$.

(2) Express P as a product of irreducible polynomials in $\mathbb{Q}[X]$, $\mathbb{R}[X]$ et $\mathbb{C}[X]$.

Solution 6.3.7.

(1) We have
$$P(1-i) = (1-i)^4 - 4(1-i)^3 + 11(1-i)^2 - 14(1-i) + 10,$$
which, upon expansion, gives $P(1-i) = 0$.

(2) Since all coefficients of $P(X)$ are real numbers, $\overline{1-i} = 1+i$ is also a root of $P(X)$. So, since
$$[X - (1-i)][X - (1+i)] = X^2 - 2X + 2,$$
and hence
$$P(X) = (X^2 - 2X + 2)Q(X).$$
To find $Q(X)$, we either use a division, or we assume that $Q(X)$ may be written as $aX^2 + bX + c$ (for $P(X)$ is of degree 4), then we find the constants a, b and c. In fact, by looking closely at $P(X)$, we see that we may even assume that $Q(X) = X^2 + bX + c$. We then have
$$(X^2 - 2X + 2)(X^2 + bX + c)$$
$$= X^4 + (b-2)X^3 + (c-2b+2)X^2 + (2b-2c)X + 2c.$$
Thus
$$P(X) = (X^2 - 2X + 2)(X^2 + bX + c) \iff \begin{cases} b - 2 = -4, \\ c - 2b + 2 = 11, \\ 2b - 2c = -14, \\ 2c = 10, \end{cases}$$
from which we derive $b = -2$ et $c = 5$. Therefore,
$$P(X) = (X^2 - 2X + 2)(X^2 - 2X + 5).$$
This is the furthest we can go in $\mathbb{Q}[X]$ and $\mathbb{R}[X]$ because the discriminants of the two polynomials on the right side are strictly negative.

In $\mathbb{C}[X]$, this is refined to:
$$P(X) = (X - 1 + i)(X - 1 - i)(X - 1 + 2i)(X - 1 - 2i).$$

Exercise 6.3.8. Find $\gcd(P(X), Q(X))$ where $P(X) = X^{99} + 1$ and $Q(X) = X^{45} + 1$.

Solution 6.3.8. Write

$$
\begin{array}{r|l}
X^{99} + 1 & X^{45} + 1 \\
\underline{-X^{99} + X^{54}} & \overline{X^{54} - X^9} \\
1 - X^{54} & \\
\underline{X^{54} + X^9} & \\
1 + X^9 & \\
\end{array}
$$

Hence

$$X^{99} + 1 = (X^{45} + 1)(X^{54} - X^9) + X^9 + 1.$$

Let us carry out another division of $X^{45} + 1$ by $X^9 + 1$. We have

$$
\begin{array}{r|l}
X^{45} + 1 & \quad X^9 + 1 \\
-X^{45} - X^{36} & X^{36} - X^{27} + X^{18} - X^9 + 1 \\
\hline
-X^{36} + 1 & \\
X^{36} + X^{27} & \\
\hline
1 + X^{27} & \\
-X^{27} - X^{18} & \\
\hline
1 - X^{18} & \\
X^{18} + X^9 & \\
\hline
1 + X^9 & \\
-X^9 - 1 & \\
\hline
0 &
\end{array}
$$

The last non-zero remainder is $X^9 + 1$, and it is the greatest common divisor. In other words,

$$\gcd(X^{99} + 1, X^{45} + 1) = X^9 + 1.$$

Exercise 6.3.9. Define over $\mathbb{R}[X]$ the following binary relation $\mathcal{R}$:

$$A\mathcal{R}B \iff A \text{ divides } B.$$

Is $\mathcal{R}$ an order relation?

Solution 6.3.9. The relation $\mathcal{R}$ is not anti-symmetric. Indeed, we have

$$X + 1 \text{ divides } 2X + 2 \text{ and } 2X + 2 \text{ divides } X + 1,$$

but $X + 1 \neq 2X + 2$.

Exercise 6.3.10. Carry out the division of the polynomial $P(X)$ by $Q(X)$ by increasing power order (up to the order 3) where

$$P(X) = 1 + X^3 + X^4 \text{ and } Q(X) = 1 + X + X^2.$$

Solution 6.3.10. We have:

$$
\begin{array}{r|l}
1 + X^3 + X^4 & 1 + X + X^2 \\
-X - X^2 + X^3 + X^4 & \quad 1 - X \\
\hline
X^4 &
\end{array}
$$

Hence

$$1 + X^3 + X^4 = (1 - X)(1 + X + X^2) + X^4$$

Exercise 6.3.11. Let
$$P(X) = X^4 + 2X^3 - 3X^2 - 4X + 4.$$

(1) Compute $P(1)$.
(2) Find $\gcd(P(X), P'(X))$.
(3) Find all real roots of $P(X)$.

Solution 6.3.11.

(1) We have
$$P(1) = 1 + 2 - 3 - 4 + 4 = 0.$$

(2) Clearly
$$P'(X) = 4X^3 + 6X^2 - 6X + 4.$$

Instead of doing a polynomial long division, we will factorize the two polynomials. We have

$$\begin{aligned}
P'(X) &= 4X^3 - 4 + 6X^2 - 6X \\
&= 4(X^3 - 1) + 6(X^2 - X) \\
&= 4(X - 1)(X^2 + X + 1) + 6X(X - 1) \\
&= 2(X - 1)(2X^2 + 5X + 2) \\
&= 2(X - 1)(X + 2)(2X + 1).
\end{aligned}$$

On the other hand, we know that $P(1) = 0$, and so $P(X)$ is divisible by $X - 1$. Readers may comfortably find that
$$P(X) = (X - 1)(X^3 + 3X^2 - 4).$$

As 1 is also a root of $X^3 + 3X^2 - 4$, we may, by one of the methods we have been using above, obtain that

$$X^3 + 3X^2 - 4 = (X - 1)(X^2 + 4X + 4) = (X - 1)(X + 2)^2,$$

so that
$$P(X) = (X - 1)^2(X + 2)^2.$$

Therefore,

$$\gcd[P(X), P'(X)] = (X - 1)(X + 2) = X^2 + X - 2.$$

(3) Obviously $P(X)$ has two (double) roots, namely: 1 and -2.

Exercise 6.3.12. Determine the partial fraction decomposition in $\mathbb{R}[X]$ of each of the following expressions:

(1) $\frac{2X}{X^2-9}$;

(2) $F(X) = \frac{3X^2+2X+1}{X(X^2+X+1)}$ (give here the decomposition in $\mathbb{C}[X]$ as well).

(3) $F(X) = \frac{2X^4+1}{X(X-1)^3(X^2+X+1)}$.

Solution 6.3.12.

(1) Let us find real a and b such that:

$$\frac{2X}{X^2-9} = \frac{a}{X-3} + \frac{b}{X+3}.$$

To find a, multiply the last equation by $X-3$. We find

$$\frac{2X}{X+3} = a + \frac{b(X-3)}{X+3}.$$

Then we let $X \to 3$, thereby $a = 1$.

As for b, multiply (still the first equation) by $X+3$ to obtain

$$\frac{2X}{X-3} = \frac{a(X+3)}{X-3} + b.$$

Then we let $X \to -3$, and we find $b = 1$. Thus,

$$\frac{2X}{X^2-9} = \frac{1}{X-3} + \frac{1}{X+3}.$$

(2) Now, we need to find real a, b and c satisfying:

$$\frac{3X^2+2X+1}{X(X^2+X+1)} = \frac{a}{X} + \frac{bX+c}{X^2+X+1}.$$

Multiplying the previous equation by X and letting $X \to 0$ yield $a = 1$. To find b et c, the safest way perhaps is to write (with $a = 1$)

$$\frac{1}{X} + \frac{bX+c}{X^2+X+1} = \frac{X^2+X+1+bX^2+cX}{X(X^2+X+1)}$$

or

$$\frac{1}{X} + \frac{bX+c}{X^2+X+1} = \frac{(1+b)X^2+(1+c)X+1}{X(X^2+X+1)}.$$

The preceding fraction is equal to $F(X)$ when and only when $b+1 = 3$ et $c+1 = 2$, i.e. $b = 2$ and $c = 1$. The required decomposition over $\mathbb{R}$ therefore is:

$$\frac{3X^2+2X+1}{X(X^2+X+1)} = \frac{1}{X} + \frac{2X+1}{X^2+X+1}.$$

In $\mathbb{C}$, $X^2 + X + 1$ admits two roots, which are $z_1 := \frac{-1-i\sqrt{3}}{2}$ and $z_2 := \frac{1+i\sqrt{3}}{2}$. Thus, the decomposition over the complex numbers is obtained from

$$\frac{3X^2 + 2X + 1}{X(X^2 + X + 1)} = \frac{a}{X} + \frac{b'}{X - z_1} + \frac{c'}{X - z_2}.$$

Obviously, the value of a obtained above need not change, i.e. $a = 1$. To find b', one multiplies the last equation by $X - z_1$, then take $X \to z_1$. We proceed in a similar way with c'. We find $b' = c' = 1$. Tus, the partial fraction decomposition in $\mathbb{C}$ is given by

$$\frac{3X^2 + 2X + 1}{X(X^2 + X + 1)} = \frac{1}{X} + \frac{1}{X - z_1} + \frac{1}{X - z_2}.$$

(3) One has to find a, b, c, d, e, and f such that

$$F(X) = \frac{a}{X} + \frac{b}{X - 1} + \frac{c}{(X - 1)^2} + \frac{d}{(X - 1)^3} + \frac{eX + f}{X^2 + X + 1}.$$

The partial fraction decomposition therefore is

$$F(X) = -\frac{1}{X} + \frac{4}{3(X - 1)} + \frac{2}{3(X - 1)^2} + \frac{1}{(X - 1)^3} - \frac{X + 1}{3(X^2 + X + 1)}.$$

Let the reader carry out the calculations.

Exercise 6.3.13. Decompose the following rational fraction in $\mathbb{R}[X]$:

$$F(X) = \frac{X^2}{(X^2 + 4)(X^2 + 1)^2}.$$

Solution 6.3.13. The procedure is the following:

$$\frac{X^2}{(X^2 + 4)(X^2 + 1)^2} = \frac{aX + b}{X^2 + 4} + \frac{cX + d}{X^2 + 1} + \frac{eX + f}{(X^2 + 1)^2}.$$

To find the real coefficient a, b, c, d, e, and f, there are at least two ways of doing so. We may gain some precious time by observing that $F(X)$ is even, and so the decomposition too must be even. Hence $a = c = e = 0$. For the other unknowns, we get $b = -\frac{4}{9}$, $d = \frac{4}{9}$, and $f = -\frac{1}{3}$. Accordingly,

$$\frac{X^2}{(X^2 + 4)(X^2 + 1)^2} = -\frac{4}{9(X^2 + 4)} + \frac{4}{9(X^2 + 1)} - \frac{1}{3(X^2 + 1)^2}.$$

6.4. Supplementary Exercises

Exercise 6.4.1. (See, e.g. [**32**]) Let F be a field and let $f(x) \in F[X]$ be a non-zero polynomial of degree n. Then $f(x)$ has at most n distinct roots in F.

Exercise 6.4.2. We define a relation $\mathcal{R}$ on $\mathbb{R}[X]$ as follows

$$P\mathcal{R}Q \iff \deg P = \deg Q.$$

Show that $\mathcal{R}$ is an equivalence relation.

Exercise 6.4.3. Decompose the following polynomials in $\mathbb{R}[X]$ and in $\mathbb{C}[X]$:
  (1) $X^3 - 5$,
  (2) $X^{12} - 1$.

Exercise 6.4.4. Find all polynomials $P \in \mathbb{C}[X]$ such that P' divides P.

Exercise 6.4.5. Let P and Q be defined by

$$P(X) = X^4 + X^3 + 2X^2 + X + 1$$

and

$$Q(X) = X^3 + 4X^2 + 4X + 3$$

respectively.
  (1) Show that

$$\gcd[P(X), Q(X)] = X^2 + X + 1.$$

  (2) Factorize $P(X)$ and $Q(X)$ in $\mathbb{R}[X]$.

Exercise 6.4.6. Let

$$P(X) = X^6 - 6X^5 + 15X^4 - 20X^3 + 12X^2 - 4.$$

  (1) Find $\gcd[P(X), P'(X)]$.
  (2) Decompose $P(X)$ in $\mathbb{R}[X]$, then in $\mathbb{C}[X]$.

Exercise 6.4.7. Decompose

$$\frac{X^3 - 3X^2 + X - 4}{X - 1}.$$

Exercise 6.4.8. Decompose

$$\frac{X^4 + 2X^2 + 1}{X^2 - 1}.$$

Exercise 6.4.9. Decompose

$$\frac{X + i}{X^2 + i}$$

in $\mathbb{C}$.

Bibliography

1. I. Adler. Composition rings, *Duke Math. J.*, **29** (1962) 607-623.
2. A. Baker. Transcendental number theory, *Cambridge University Press, London-New York*, 1975.
3. M. Bhargava. Groups as unions of proper subgroups, *Amer. Math. Monthly*, **116/5** (2009) 413-422.
4. Bibmath, https://www.bibmath.net/
5. G. Birkhoff, S. Mac Lane. A brief survey of modern algebra. Second edition. *The Macmillan Co., New York; Collier-Macmillan Ltd., London* 1965.
6. S. M. Buckley, D. MacHale. Variations on a theme: rings satisfying $x^3 = x$ are commutative, *Amer. Math. Monthly*, **120/5** (2013) 430-440.
7. P. J. Cameron, Introduction to Algebra, *Oxford Science Publications*, 1998.
8. R. Coleman. Some properties of finite rings, arXiv preprint (2013). arXiv: 1302.3192.
9. J. d'Angelo, D. West. *Mathematical Thinking, Problem-Solving and Proofs.* Prentice Hall, Englewood Cliffs, N.J., 2nd edition, 2000.
10. S. Dehimi, M. H. Mortad. Unbounded operators having self-adjoint or normal powers and some related results, (submitted). arXiv:2007.14349.
11. A. Denmat, F. Héaulme, *Algèbre Générale* (French), Série: TD, Dunod 2000.
12. D. S. Dummit, R. M. Foote. Abstract algebra. Third edition. *John Wiley & Sons, Inc., Hoboken, NJ*, 2004.
13. H. Gianella, R. Krust, F. Taieb, N. Tosel. Problèmes Choisis de Mathématiques Supérieures (French), *Springer*, 2001.
14. E. Hecke. Lectures on the theory of algebraic numbers. Translated from the German by George U. Brauer, Jay R. Goldman and R. Kotzen. Graduate Texts in Mathematics, **77**. *Springer-Verlag, New York-Berlin*, 1981.
15. I. N. Herstein. Topics in algebra, Second edition. *Xerox College Publishing, Lexington, Mass.-Toronto, Ont.*, 1975.
16. E. Hille. Gelfond's solution of Hilbert's seventh problem, *Amer. Math. Monthly*, **49** (1942) 654-661.
17. J. M. Howie. Complex analysis. Springer Undergraduate Mathematics Series. *Springer-Verlag London, Ltd., London*, 2003.
18. M. Kuzucuoğlu, Exercises and Solutions in Groups Rings and Fields. 2012. http://users.metu.edu.tr/matmah/Graduate-Algebra-Solutions/ Undergraduate-Algebra-Problems%20and%20Solutions.pdf
19. E. Lamb. Two-hundred-terabyte maths proof is largest ever, *Nature*, **534** (2016) 17-18. https://doi.org/10.1038/nature.2016.19990.
20. F. M. S. Lima. Some transcendence results from a harmless irrationality theorem, *J. Ana. Num. Theor.*, **5/2** (2017) 91-96.

21. S. Lipschutz, *Schaum's Outline of Set Theory and Related Topics*, McGraw-Hill, (2nd edition) 1998.

22. M. H. Mortad. *Introductory topology. Exercises and solutions.* 2nd edition. (English). Hackensack, NJ: World Scientific. (2017).

23. M. H. Mortad. *An operator theory problem book*, World Scientific Publishing Co., (2018). https://doi.org/10.1142/10884. ISBN: 978-981-3236-25-7 (hardcover).

24. M. H. Mortad. On the existence of normal square and nth roots of operators, *J. Anal.*, **28/3** (2020) 695-703.

25. M. H. Mortad. Basic Linear Algebra: Exercises and Solutions, World Scientific Publishing Co., (to appear).

26. M. H. Mortad. Basic Real Analysis: Exercises and Solutions, World Scientific Publishing Co., (to appear).

27. M. H. Mortad. *Counterexamples in operator theory*, (submitted book).

28. M. I. Rosen. Niels Hendrik Abel and equations of the fifth degree, *Amer. Math. Monthly*, **102/6** (1995) 495-505.

29. R. E. Schwartz, http://www.math.brown.edu/reschwar/MathNotes/notes5.pdf

30. M. Spivak. Calculus. Publish or Perish, Inc. Houston, Third Edition 1994.

31. M. Stoll, *Real Analysis*, Addison-Wesley Higher Mathematics, Second Edition 2001.

32. C. Whitehead, *Guide² Algebra*, Palgrave Mathematical Guides, Second Edition 2003.

Index

Printed in the United States
by Baker & Taylor Publisher Services